Ion Implantation and Activation

(Volume 3)

Authored By

Kunihiro Suzuki

Fujitsu limited
Minatoku kaigan 1-11-1 Tokyo
Japan

CONTENTS

PREFACE

Ion implantation is a standard technology, by which impurities are doped into Si substrates in very large scale integration (VLSI) processes in modern times. Defects are unintentionally generated in the substrate by the ion implantation simultaneously since impurities are introduced physically into the lattice, *i.e.,* in non-thermal equilibrium. Hence, a subsequent thermal process is indispensable to recover the damage and activate the introduced impurities. Intensive studies on predicting the ion implantation profiles and activation processes have been done, and they are still under investigations.

Many books on fundamental ion implantation technology and detailed physics related to the ion implantation have been published. Furthermore, database for practical ion implantation profiles is also established and it is valid even for modern technologists and students. However, theoretical models and systematic fundamental experimental data are not closely related to each other, and how accurately the models reproduce the experimental data is not clear.

We have been collecting systematic experimental data to evaluate and develop the models for a long time. We also try to compare the data with related models in this eBook.

Details about the derivation process of fundamental and advanced models are described step by step and their related assumptions and approximations are clarified. If some mathematical and physical background is necessary, related appendixes will be added to make this book self-contained. Models which are not established well and under investigations are also treated. Although it is not ensured to be accurate, knowing the detail about models under investigations is interesting for furthering the models.

The author aims at the readers who are experts and non-experts in various fields associated with the ion implantation, hoping that various members can cover some knowledge to collaborate with one another. It is not easy for non-expert members to understand this eBook, it is believed that they can do it with time and efforts.

This eBook consists of three volumes, and this volume treats the following content.

The thermal process after the ion implantation is indispensable to activate the introduced impurities. The impurities diffuse and redistribute in the thermal process. It is explained that the diffusion phenomenon is associated with the paring of impurities and point defects. Firstly, we describe the diffusion phenomenon under the thermal equilibrium on point defect concentration.

The ion implantation unintentionally induces point defects much more than that in the thermal equilibrium, and induces significant diffusion much more than that in the thermal equilibrium. This phenomenon is called transient enhanced diffusion (TED) and is not fully understood. The prominent feature of TED is shown and described.

S_iO_2 is a very important material in VLI process, which not only provides good interface characteristics in MOSFETs but also blocks the impurity diffusion. SiO_2 layers are easily formed by thermal oxidation. Mechanisms of oxidation and redistribution of impurities during oxidation are presented.

Segregaton coefficient is defined as the ratio of impurity concentrations at both sides of two different layers, and it influences the diffusion profiles in multi layers. The most important segregation coefficient is the one for SiO_2/Si system. Although the definition of the segregation coefficient is simple, its evaluation is not easy. We show that the oxidation of polycrystalline silicon enables us to extract it robustly.

The diffusion equation cannot be solved analytically in general. However, we can obtain their analytical forms in some special cases. We describe some analytical models related to the VLSI process.

CONFLICT OF INTEREST

The author(s) confirm that this chapter content has no conflict of interest.

Kunihiro Suzuki

Fujitsu limited
Minatoku kaigan 1-11-1 Tokyo
Japan
E-mail: suzuki.kunihiro@jp.fujitsu.com

ACKNOWLEDGEMENTS

Many members in various fields contributed to complete this eBook.

Tsutomu Nagayama helped me to obtain many ion implantation samples with ion implantation machine, and Susumu Nagayama helped me to obtain many SIMS data. Yasuharu Fujimori helped me to obtain various carrier profiles.

Masatoshi Yoshihara and Syuichi Kojima contributed to integrate models into a system which is used by many members.

I thank Prof. Fichtner for inviting me to his laboratory, where I started studying process modeling and enjoyed the collaboration with Dr. Alexander Hoefler, Dr. Thomas Feudel, and Dr. Nobert Strecker.

Dr. Christoph Zechner gave cutting edge information on process modeling, and I enjoyed some collaboration with him in the field. He also gave me invaluable discussions, comments, and corrected my English.

Hiroko Tashiro and Ritsuo Sudo helped me to simulate ion implantation and diffusion profiles.

Hiroyo Miyamoto helped me to set format of word file.

I owe to Dr. Min Yu a great deal for his invaluable discussions and correcting my English. His devotion made it possible to complete this eBook.

Prof. Seijiro Furukawa and Prof. Hiroshi Ishihara directed me to become a researcher.

I also want to give special thanks to Prof. Robert W. Dutton who always encouraged me from the beginning of my career as a researcher.

Finally, I want to thank my family Kyoko, Yuji, Saori, Takayuki, Misato, Nahoko, Ryohei, and my parents Fukushi and Yae, and Kyoko's ex-father Haruo Abe, and mother Chiyoko Abe with whom I enjoy comfortable life.

Kunihiro Suzuki

Fujitsu limited
Minatoku kaigan 1-11-1 Tokyo
Japan

Ion Implantation and Activation

(Volume 3)

2

Send Orders for Reprints to reprints@benthamscience.net
Ion Implantation and Activation, Vol. 3, 2013, 3-34

CHAPTER 1

Diffusion Under Thermal Equilibrium

Abstract: We derive a diffusion flux and diffusion equation step by step, starting with a rough sketch one to a practical one. The flux is described with the paring with point defects, and the dependence of diffusion coefficients on doping concentration is emerged. We can predict diffusion profiles by solving the equation.

Keywords: Ion implantation, diffusion, diffusion flux, diffusion equation, diffusion coefficient, point defects, doping concentration, lattice sites, jumping frequency, electric field, mobility, Einstein's relationship, interstitial Si, intrinsic carrier concentration, statistic mechanism.

INTRODUCTION

Ion implanted impurities are introduced to Si substrates physically, and they are not set at the lattice sites in general. Further, the impurities introduce many defects, and they do not play a role of donors or acceptors as they are. Thermal processes, that are, annealing processes are indispensable to activate the impurities. The impurities redistribute during these annealing processes, and the final redistributed profiles determine the device characteristics. Therefore, it is important to predict the diffusion profiles.

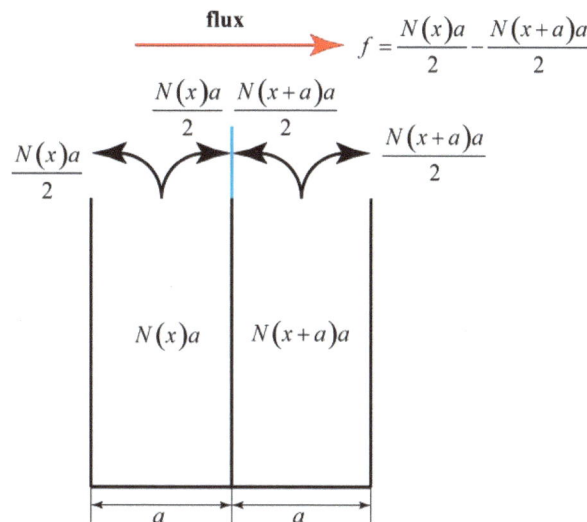

$$f = \frac{N(x)a}{2} - \frac{N(x+a)a}{2}$$

Figure 1: Schematic diffusion flux.

Kunihiro Suzuki

DIFFUSION FLUX

The flux f of any species is defined as the number of particles passing through unit area in unit time. Let us consider two boxes of length a as shown in Fig. **1**. The location of the left side box is x, and the corresponding concentration is $N(x)$. The location of the right side box is $x+a$, and the corresponding concentration is $N(x+a)$. Therefore, the amount of impurities in the boxes are $N(x)a$ and $N(x+a)a$, respectively. We assume that impurities in the boxes randomly jump to the left or right box in the next step. Therefore, half of the impurities go to the right box, and the others to the left one. The flux from the x-location box to the $x+a$-location box f_1 is then expressed by

$$f_1 = \frac{1}{2}N(x)a \tag{1}$$

On the other hand, the flux from $x+a$-location box to the x-location box f_2 is given by

$$f_2 = \frac{1}{2}N(x+a)a \tag{2}$$

Consequently, the net flux that crosses the boundary of the both boxes f is given by

$$
\begin{aligned}
f &= f_1 - f_2 \\
&= \frac{1}{2}N(x)a - \frac{1}{2}N(x+a)a \\
&\approx \frac{1}{2}N(x)a - \frac{1}{2}\left[N(x) + \frac{\partial N}{\partial x}a\right]a \\
&= -\frac{1}{2}a^2\frac{\partial N}{\partial x}
\end{aligned}
\tag{3}
$$

Note that the flux is proportional to the gradient of the concentration. This is attributed to the random motion of the impurities.

Some points are oversimplified in the above discussion.

We assumed that all impurities move to the left or the right box. However, the crystal substrate atoms form a series of potentials hills which impede the impurities to move to other boxes as shown in Fig. **2**. The impurities are not in the box, but in the valley of potential with a barrier height of ΔE. The impurities should overcome the potential barrier ΔE, and most of the impurities stay at their original location. According to the statistic mechanism, the probability of the impurities that overcome the potential hill, and move to the neighbor location is proportional to

$$\exp\left(-\frac{\Delta E}{k_B T}\right)$$

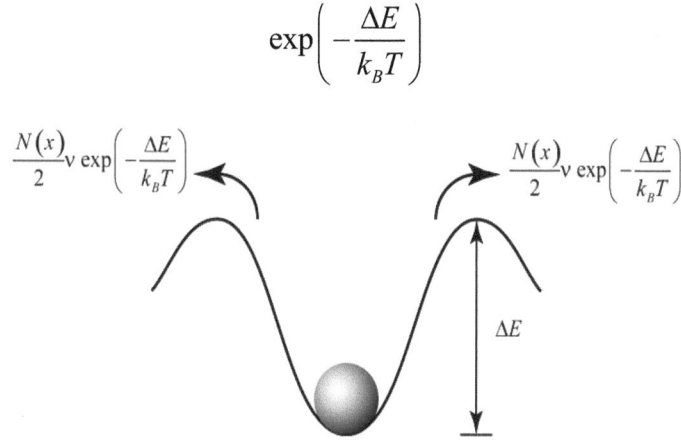

$$\frac{N(x)}{2}v\exp\left(-\frac{\Delta E}{k_B T}\right) \qquad \frac{N(x)}{2}v\exp\left(-\frac{\Delta E}{k_B T}\right)$$

$$\Delta E$$

Figure 2: An impurity in diffusion potential.

We should multiply this probability to the flux. Further, we should consider the factor of time, that is, attempt frequency v. This expresses how many times each impurity tries to overcome the potential hill. Finally each flux is given by

$$f_1 = \frac{1}{2}v\exp\left(-\frac{\Delta E}{k_B T}\right)N(x)a \qquad (4)$$

$$f_2 = \frac{1}{2}v\exp\left(-\frac{\Delta E}{k_B T}\right)N(x+a)a \qquad (5)$$

The net flux f is then given by

$$f = -\frac{1}{2}v\exp\left(-\frac{\Delta E}{k_B T}\right)a^2\frac{\partial N}{\partial x}$$

$$= -D\frac{\partial N}{\partial x} \qquad (6)$$

where D is the diffusion coefficient and has a general form of

$$D = D_0 \exp\left(-\frac{\Delta E}{k_B T}\right) \tag{7}$$

Comparing Eq. 6 with Eq. 7, D_0 is given by

$$D_0 = \frac{1}{2}\nu a^2 \tag{8}$$

Let us consider the jumping frequency ν. We expand the diffusion potential into Taylor series with respect to the minimum point x_0 as

$$
\begin{aligned}
V(x) &\approx V(x_0) + (x - x_0)\frac{dV(x)}{dx}\bigg|_{x=x_0} + \frac{1}{2}(x - x_0)^2 \frac{d^2V(x)}{dx^2}\bigg|_{x=x_0} \\
&= V(x_0) + \frac{1}{2}(x - x_0)^2 \frac{d^2V(x)}{dx^2}\bigg|_{x=x_0} \\
&= V(x_0) + \frac{1}{2}k(x - x_0)^2
\end{aligned} \tag{9}
$$

where

$$k \equiv \frac{d^2V(x)}{dx^2}\bigg|_{x=x_0} \tag{10}$$

The force associated with the potential can be expressed by

$$F = -\frac{dV}{dx} = -k(x - x_0) \tag{11}$$

The corresponding Newton equation is given by

$$m\frac{\partial^2(x - x_0)}{\partial t^2} = -k(x - x_0) \tag{12}$$

This is solved as

$$x - x_0 = A\cos(\omega t + \delta) \tag{13}$$

where

$$\omega = \sqrt{\frac{k}{m}} \tag{14}$$

δ is the initial phase, which is not important for the final result. The jumping frequency ν is then given by

$$\nu = \frac{\omega}{2\pi}\sqrt{\frac{k}{m}} = \frac{\omega}{2\pi}\sqrt{\frac{\left.\dfrac{d^2V(x)}{dx^2}\right|_{x=x_0}}{m}} \tag{15}$$

ν is determined by the shape of the potential. The substrate lattice distance is almost invariable even when the temperature changes. Therefore, the potential shape is rather independent of the temperature, and is thought to be of the order of $10^{12}/\sec$ [1].

The diffusion barrier is thought to be of the order of 0.3 V [2], and jumping distance is about the Si lattice spacing of 0.565 nm, and use mass of a B. Substituting these values to Eq. 9, we obtain k as

$$\frac{1}{2}k\left(\frac{1}{2}\times 0.565\times 10^{-9}\right)^2 = 0.3\times 1.6\times 10^{-19}$$

$$k = \frac{2\times 0.3\times 1.6\times 10^{-19}}{\left(\dfrac{1}{2}\times 0.565\times 10^{-9}\right)^2} = 1.2$$

(Equation 9 is only valid near the bottom of potential, but we assume that is valid for the distance of $a\!/_2$ for simplicity.) Therefore, we obtain

$$\nu = \frac{1}{2\pi}\sqrt{\frac{k}{m}}$$

$$= \frac{1}{2\pi}\sqrt{\frac{1.2}{\dfrac{10.8\times 10^{-3}}{6.03\times 10^{23}}}}$$

$$= 1.3\times 10^{12}\ [/s]$$

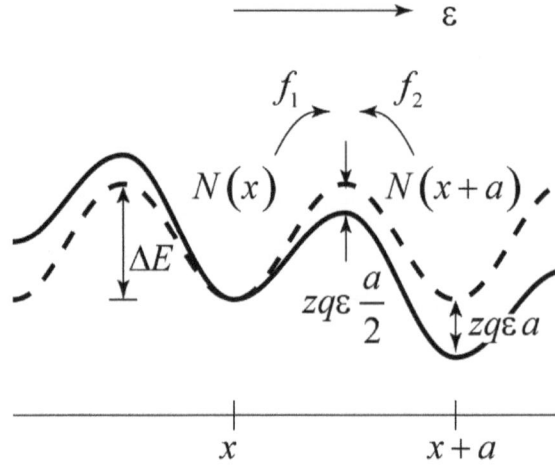

Figure 3: Diffusion potential under electric field.

In the so far derivation, we do not apply external electric fields during the diffusion process, but only built in electric field ε due to non-uniform doping profile. If the diffusion species are charged with z, we should consider a drift component.

The corresponding diffusion potential under electric field is then tilted as shown in Fig. **3**, where a dashed line corresponds to the one without electric field. The potential from the location x to $x+a$ is lowered by $zq\varepsilon\%_2$, and that from $x+a$ to x by $zq\varepsilon a$. Therefore, f_1 is enhanced by the electric field. Each flux is then given by

$$
\begin{aligned}
f_1 &= \frac{1}{2}vaN(x)\exp\left(-\frac{\Delta E - zq\frac{1}{2}a\varepsilon}{k_BT}\right) \\
&\approx \frac{1}{2}vaN(x)\exp\left(-\frac{\Delta E}{k_BT}\right)\left(1+\frac{zq\frac{1}{2}a\varepsilon}{k_BT}\right) \\
&= \frac{1}{2}vaN(x)\exp\left(-\frac{\Delta E}{k_BT}\right)+\frac{1}{4}z\frac{q}{k_BT}va^2\exp\left(-\frac{\Delta E}{k_BT}\right)\varepsilon N(x)
\end{aligned}
\tag{16}
$$

$$f_2 = \frac{1}{2} v a N (x+a) \exp\left(-\frac{\Delta E + z q \frac{1}{2} a \varepsilon}{k_B T} \right)$$

$$\approx \frac{1}{2} v a \left[N(x) + \frac{\partial N(x)}{\partial x} a \right] \exp\left(-\frac{\Delta E}{k_B T} \right) \left(1 - \frac{z q \frac{1}{2} a \varepsilon}{k_B T} \right) \qquad (17)$$

$$\approx \frac{1}{2} v a N(x) \exp\left(-\frac{\Delta E}{k_B T} \right) - \frac{1}{4} z \frac{q}{k_B T} v a^2 \exp\left(-\frac{\Delta E}{k_B T} \right) \varepsilon N(x)$$

$$+ \frac{1}{2} v a^2 \exp\left(-\frac{\Delta E}{k_B T} \right) \frac{\partial N(x)}{\partial x}$$

Therefore, the net flux is given by

$$f = f_1 - f_2$$

$$= -\frac{1}{2} v a^2 \exp\left(-\frac{\Delta E}{k_B T} \right) \frac{\partial N(x)}{\partial x} + z \frac{q}{k_B T} \frac{1}{2} v a^2 \exp\left(-\frac{\Delta E}{k_B T} \right) \varepsilon N(x) \qquad (18)$$

$$= -D \frac{\partial N(x)}{\partial x} + z \mu \varepsilon N(x)$$

where D is the diffusion coefficient, and μ is the mobility.

$$D \equiv \frac{1}{2} v a^2 \exp\left(-\frac{\Delta E}{k_B T} \right), \mu \equiv \frac{q}{k_B T} \frac{1}{2} v a^2 \exp\left(-\frac{\Delta E}{k_B T} \right) \qquad (19)$$

Note that the diffusion coefficient and mobility are related by

$$D = \frac{k_B T}{q} \mu \qquad (20)$$

This is called as Einstein's relationship.

DIFFUSION FLUX ASSOCIATED WITH POINT DEFECTS

The flux was derived assuming that diffusion species are impurities in the previous section. However, impurities are believed to diffuse not by themselves only, but pairing with point defects. The detailed mechanism is still under discussion. The rough sketch of diffusion is as shown in Fig. **4**.

a) Impurity does not diffuse itself and stays at the lattice site (Fig. **4**(a)).

b) Interstitial Si diffuse freely in the Si lattice and one of them is close to a lattice site impurity by chance (Fig. **4**(b)).

c) The interstitial Si and impurity form a pair (Fig. **4**(c)), and it has a chance to move to one of the neighbor four lattice sites.

d) The impurity exchanges the pair Si and move to its location. The impurity moves down in Fig. **4**(d).

e) The impurity exchanges the pair Si and move to its location. The impurity changes its pair and move to the left in Fig. **4**(e).

f) The impurity dissolves and impurity set at the lattice site releasing aninterstitial Si in Fig. **4**(f).

The corresponding diffusion potential is shown in Fig. **5** [2].

Consider the reaction between z_j charged impurities $A_j^{(z_j)}$ and z charged interstitial Si $I^{(z)}$ which is given by

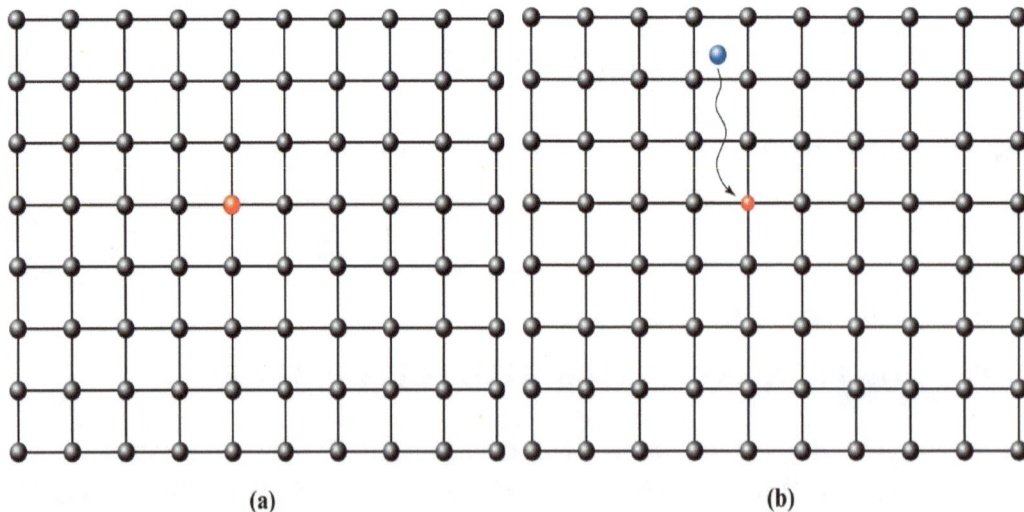

$$A_j^{(z_j)} + I^{(z)} \underset{k_2}{\overset{k_1}{\rightleftarrows}} \left(A_j I\right)^{(z_j+z)} \tag{21}$$

(a)   (b)

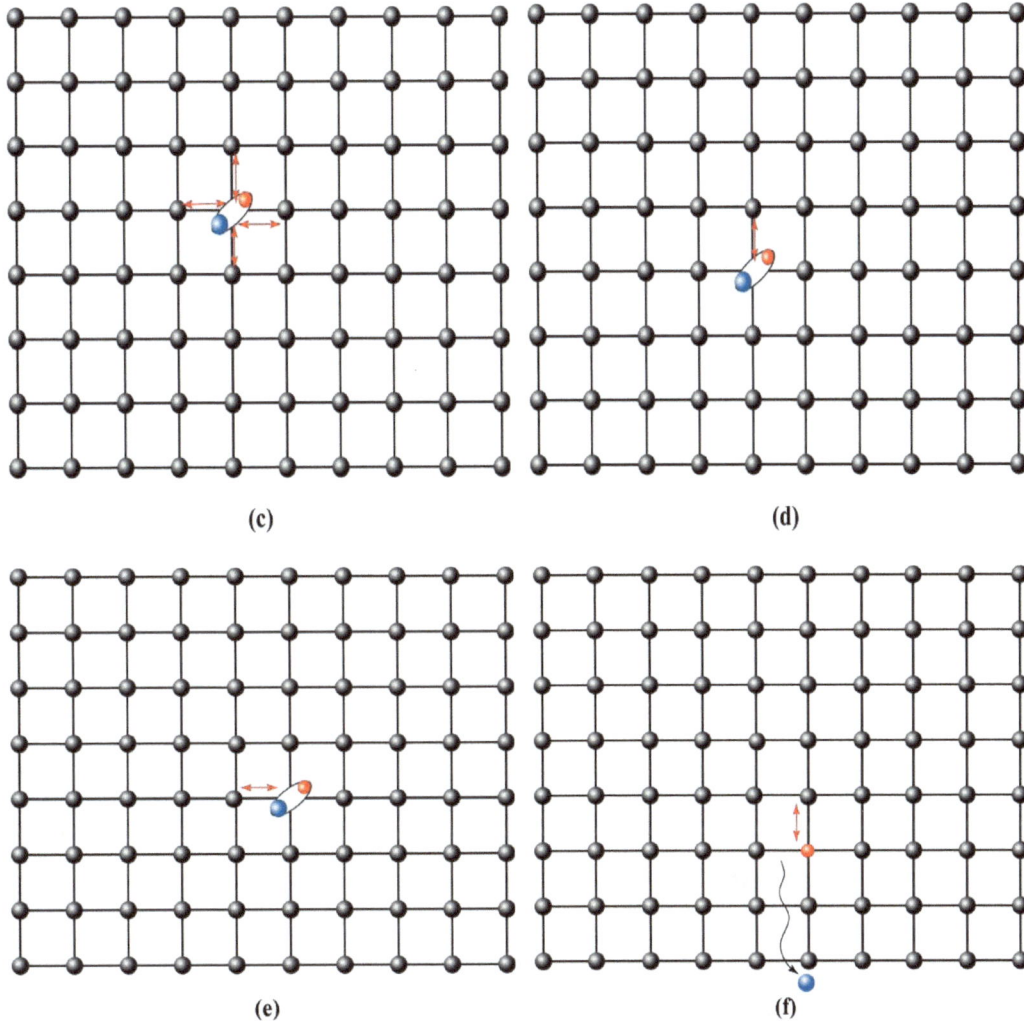

Figure 4: Diffusion animation associated with interstitial Si paring mechanism. (a) Impurity stays at the lattice site. (b) Diffused Interstitial Si is close to a lattice site impurity by chance. (c) The interstitial Si and impurity form a pair, which have a chance to move to four directions. (d) The impurity exchanges the pair Si and movesdown. (e) The impurity exchanges the pair Si and movesto right. (f) The pairdissolves and impurity set at the lattice site releasing aninterstitial Si.

The flux of $\left(A_j I\right)^{\left(z_j + z\right)}$ related to k_1 in Eq. 21 is expressed by

$$\frac{\partial \left| \left(A_j I\right)^{\left(z_j + z\right)} \right|}{\partial t} = \frac{1}{2} \left[A_j^{(z_j)} \right] \frac{\left[I^{(z)} \right]}{\left[Si \right]} v_1 \exp\left(-\frac{\Delta E_1}{k_B T} \right) \qquad (22)$$

k₁:B kick out   **k₂:B kick in**

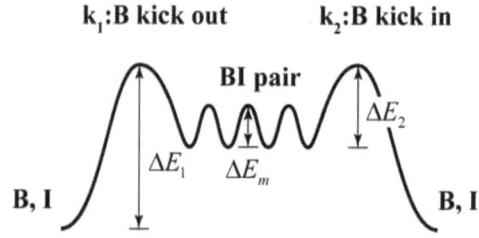

Figure 5: Diffusion potential considering impurity point defect pair [2]. We assume impurity of B here, but it is accommodate for the other impurities with different parameters.

The flux of $\left(A_j I \right)^{(z_j + z)}$ related to k_2 is expressed by

$$\frac{\partial \left[\left(A_j I \right)^{(z_j + z)} \right]}{\partial t} = -\frac{1}{2} \left[\left(A_j I \right)^{(z_j + z)} \right] \nu_2 \exp\left(-\frac{\Delta E_2}{k_B T} \right) \tag{23}$$

Assuming that the thermal equilibrium of the reaction is established, we obtain

$$\left[\left(A_j I \right)^{(z_j + z)} \right] = \frac{\nu_1}{\nu_2} \left[A_j^{(z_j)} \right] \frac{\left[I^{(z)} \right]}{[Si]} \nu_1 \exp\left[-\frac{\left(\Delta E_1 - \Delta E_2 \right)}{k_B T} \right] \tag{24}$$

The generation energy of neutral interstitial Si is assumed to be ΔE_I, and the corresponding concentration is expressed by

$$\left[I^{(0)} \right] = r_v [Si] \exp\left(-\frac{\Delta E_I}{k_B T} \right) \tag{25}$$

where r_v is the ratio of jumping frequency of interstitial Si to that of lattice Si. Further, the charging reaction of neutral Si to z charged one is expressed by

$$I^{(0)} + zh \rightleftarrows I^{(z)} \tag{26}$$

We assume the reaction coefficient of K is constant independent of carrier concentration, and using one with general and intrinsic semiconductor, and we obtain

$$K = \frac{\left[I^{(z)} \right]}{\left[I^{(0)} \right] p^z} = \frac{\left[I^{(z)} \right]_i}{\left[I^{(0)} \right]_i n_i^z} \tag{27}$$

We then obtain

$$
\begin{aligned}
\left[I^{(z)} \right] &= \frac{\left[I^{(z)} \right]_i}{\left[I^{(0)} \right]_i} \left[I^{(0)} \right] \left(\frac{p}{n_i} \right)^z \\
&= \frac{\left[I^{(z)} \right]_i}{\left[I^{(0)} \right]_i} \left[I^{(0)} \right] \left(\frac{n}{n_i} \right)^{-z} \\
&= \frac{\left[I^{(z)} \right]_i}{\left[I^{(0)} \right]_i} \left(\frac{n}{n_i} \right)^{-z} r_v \left[Si \right] \exp \left(-\frac{\Delta E_I}{k_B T} \right)
\end{aligned}
\tag{28}
$$

Therefore, we obtain

$$
\left[\left(A_j I \right)^{(z_j + z)} \right] = r_v \frac{v_1}{v_2} \left[A_j^{(z_j)} \right] \frac{\left[I^{(z)} \right]_i}{\left[I^{(0)} \right]_i} \left(\frac{n}{n_i} \right)^{-z} \exp \left[-\frac{\Delta E_I + \left(\Delta E_1 - \Delta E_2 \right)}{k_B T} \right]
\tag{29}
$$

We further express

$$
\left[I^{(z)} \right]_i = \left[I^{(0)} \right]_i \exp \left(-\frac{\Delta E_z}{k_B T} \right)
\tag{30}
$$

and obtain

$$
\left[\left(A_j I \right)^{(z_j + z)} \right] = r_v \frac{v_1}{v_2} \left[A_j^{(z_j)} \right] \left(\frac{n}{n_i} \right)^{-z} \exp \left[-\frac{\Delta E_I + \Delta E_z + \left(\Delta E_1 - \Delta E_2 \right)}{k_B T} \right]
\tag{31}
$$

$\left(A_j I \right)^{(z_j + z)}$ is a real diffusion species, and its flux is expressed by

$$
\begin{aligned}
f &= -\frac{1}{2} v_m a^2 \exp \left[-\frac{\Delta E_m}{k_B T} \right] \frac{\partial \left[\left(A_j I \right)^{(z_j + z)} \right]}{\partial x} \\
&= -D_{A_j I} \frac{\partial \left[\left(A_j I \right)^{(z_j + z)} \right]}{\partial x}
\end{aligned}
\tag{32}
$$

where

$$D_{A_jI} \equiv \frac{1}{2} v_m a^2 \exp\left(-\frac{\Delta E_m}{k_B T}\right) \tag{33}$$

This species are $z_j + z$ charged and we should add a drift component as

$$f = -D_{A_jI} \frac{\partial\left[\left(A_j I\right)^{\left(z_j+z\right)}\right]}{\partial x} + \left(z_j + z\right)\mu\varepsilon\left[\left(A_j I\right)^{\left(z_j+z\right)}\right] \tag{34}$$

where the electric field ε is expressed as follows. The electron concentration n is expressed by the electric potential as

$$n = n_i \exp\left(\frac{q\phi}{k_B T}\right) \tag{35}$$

The electric field is then given by

$$\varepsilon = -\frac{\partial\phi}{\partial x} = -\frac{k_B T}{q} \frac{\partial \ln\left(\dfrac{n}{n_i}\right)}{\partial x} \tag{36}$$

Substituting Eq. 36 into Eq. 34, we obtain

$$
\begin{aligned}
f &= -D_{A_jI} \frac{\partial\left[\left(A_j I\right)^{\left(z_j+z\right)}\right]}{\partial x} - \left(z_j+z\right)\frac{k_B T}{q}\mu\frac{\partial \ln\left(\dfrac{n}{n_i}\right)}{\partial x}\left[\left(A_j I\right)^{\left(z_j+z\right)}\right] \\
&= -D_{A_jI} \frac{\partial\left[\left(A_j I\right)^{\left(z_j+z\right)}\right]}{\partial x} - D_{A_jI}\frac{\partial \ln\left(\dfrac{n}{n_i}\right)^{\left(z_j+z\right)}}{\partial x}\left[\left(A_j I\right)^{\left(z_j+z\right)}\right] \\
&= -D_{A_jI}\left\{ \frac{\partial\left[\left(A_j I\right)^{\left(z_j+z\right)}\right]}{\partial x} + \left(\frac{n}{n_i}\right)^{-\left(z_j+z\right)}\frac{\partial \left(\dfrac{n}{n_i}\right)^{\left(z_j+z\right)}}{\partial x}\left[\left(A_j I\right)^{\left(z_j+z\right)}\right] \right\}
\end{aligned}
$$

$$= -D_{A_jI}\left(\frac{n}{n_i}\right)^{-(z_j+z)}\left\{\left(\frac{n}{n_i}\right)^{(z_j+z)}\frac{\partial\left[\left(A_jI\right)^{(z_j+z)}\right]}{\partial x}+\frac{\partial\left(\frac{n}{n_i}\right)^{(z_j+z)}}{\partial x}\left[\left(A_jI\right)^{(z_j+z)}\right]\right\}$$

$$= -D_{A_jI}\left(\frac{n}{n_i}\right)^{-(z_j+z)}\frac{\partial}{\partial x}\left\{\left(\frac{n}{n_i}\right)^{(z_j+z)}\left(A_jI\right)^{(z_j+z)}\right\} \tag{37}$$

$$= -\left(\frac{n}{n_i}\right)^{-z}D_{A_jI}\left(\frac{n}{n_i}\right)^{-z_j}\frac{\partial}{\partial x}\left\{\left(\frac{n}{n_i}\right)^{(z_j+z)}\left(A_jI\right)^{(z_j+z)}\right\}$$

We observe $\left|A_j^{(z_j)}\right|$ distribution experimentally instead of A_jI pair experimentally, and hence convert the flux to the corresponding one as

$$f = -\left(\frac{n}{n_i}\right)^{-z}D_{A_jI}\left(\frac{n}{n_i}\right)^{-z_j}\frac{\partial}{\partial x}\left\{\left(\frac{n}{n_i}\right)^{(z_j+z)}r_v\frac{v_1}{v_2}\left[A_j^{(z_j)}\right]\left(\frac{n}{n_i}\right)^{-z}\exp\left[-\frac{\Delta E_I+\Delta E_z+\left(\Delta E_1-\Delta E_2\right)}{k_BT}\right]\right\}$$

$$= -\left(\frac{n}{n_i}\right)^{-z}D_{A_jI}r_v\frac{v_1}{v_2}\exp\left[-\frac{\Delta E_I+\Delta E_z+\left(\Delta E_1-\Delta E_2\right)}{k_BT}\right]\left(\frac{n}{n_i}\right)^{-z_j}\frac{\partial}{\partial x}\left\{\left(\frac{n}{n_i}\right)^{z_j}\left[A_j^{(z_j)}\right]\right\} \tag{38}$$

$$= -\left(\frac{n}{n_i}\right)^{-z}D_{A_j\_z}\left(\frac{n}{n_i}\right)^{-z_j}\frac{\partial}{\partial x}\left\{\left(\frac{n}{n_i}\right)^{z_j}\left[A_j^{(z_j)}\right]\right\}$$

where

$$D_{A_j\_z} = D_{A_jI}r_v\frac{v_1}{v_2}\exp\left[-\frac{\Delta E_I+\Delta E_z+\left(\Delta E_1-\Delta E_2\right)}{k_BT}\right]$$

$$= \frac{1}{2}v_ma^2r_v\frac{v_1}{v_2}\exp\left[-\frac{\Delta E_m}{k_BT}\right]\exp\left[-\frac{\Delta E_I+\Delta E_z+\left(\Delta E_1-\Delta E_2\right)}{k_BT}\right] \tag{39}$$

$$= \frac{1}{2}v_ma^2r_v\frac{v_1}{v_2}\exp\left[-\frac{\Delta E_m+\Delta E_I+\Delta E_z+\left(\Delta E_1-\Delta E_2\right)}{k_BT}\right]$$

We consider the relationship between n and N. The impurity concentration is related to electron and whole concentration n, p as

$$N \equiv N_D - N_A = n - p \tag{40}$$

where

$$np = n_i^2 \tag{41}$$

is held. We then obtain

$$\frac{n}{n_i} = \frac{N + \sqrt{N^2 + 4n_i^2}}{2n_i} \tag{42}$$

When the impurity is a donor, $z_j = +1$, $N = \lfloor A_j^{(+)} \rfloor$.

$$\left(\frac{n}{n_i}\right)^{-z_j} \frac{\partial}{\partial x} \left\{ \left(\frac{n}{n_i}\right)^{z_j} \left[A_j^{(z_j)}\right] \right\}$$

$$= \left(\frac{n}{n_i}\right)^{-1} \frac{\partial}{\partial x} \left\{ \left(\frac{n}{n_i}\right) \left[A_j^{(z_j)}\right] \right\}$$

$$= \frac{\partial \left[A_j^{(z_j)}\right]}{\partial x} + \left[A_j^{(z_j)}\right] \left(\frac{n}{n_i}\right)^{-1} \frac{\partial}{\partial x}\left(\frac{n}{n_i}\right)$$

$$= \frac{\partial \left[A_j^{(z_j)}\right]}{\partial x} + \left[A_j^{(z_j)}\right] \frac{1}{\left[A_j^{(z_j)}\right] + \sqrt{\left[A_j^{(z_j)}\right]^2 + 4n_i^2}} \left(1 + \frac{\left[A_j^{(z_j)}\right]}{\sqrt{\left[A_j^{(z_j)}\right]^2 + 4n_i^2}}\right) \frac{\partial \left[A_j^{(z_j)}\right]}{\partial x}$$

$$= \left[1 + \frac{1}{\sqrt{1 + \left(\dfrac{4n_i}{\left[A_j^{(z_j)}\right]}\right)^2}}\right] \frac{\partial \left[A_j^{(z_j)}\right]}{\partial x} \tag{43}$$

Therefore, the flux is given by

$$
f = -\left(\frac{n}{n_i}\right)^{-z} D_{A_j{-}z}\left(\frac{n}{n_i}\right)^{-z_j}\frac{\partial}{\partial x}\left\{\left(\frac{n}{n_i}\right)^{z_j}\left[A_j^{(z_j)}\right]\right\}
$$

$$
= -\left(\frac{n}{n_i}\right)^{-z} D_{A_j{-}z}\left[1 + \frac{1}{\sqrt{1 + \left(\dfrac{4n_i}{\left[A_j^{(z_j)}\right]}\right)^2}}\right]\frac{\partial\left[A_j^{(z_j)}\right]}{\partial x}
\tag{44}
$$

When the impurity is an acceptor, $z_j = -1$, $N = -\left\lfloor A_j^{(+)}\right\rfloor$. Then, the flux is

$$
\left(\frac{n}{n_i}\right)^{-z_j}\frac{\partial}{\partial x}\left\{\left(\frac{n}{n_i}\right)^{z_j}\left[A_j^{(z_j)}\right]\right\}
$$

$$
= \left(\frac{n}{n_i}\right)\frac{\partial}{\partial x}\left\{\left(\frac{n}{n_i}\right)^{-1}\left[A_j^{(z_j)}\right]\right\}
$$

$$
= \frac{\partial\left[A_j^{(z_j)}\right]}{\partial x} + \left[A_j^{(z_j)}\right]\left(\frac{n}{n_i}\right)\frac{\partial}{\partial x}\left(\frac{n}{n_i}\right)^{-1}
$$

$$
= \frac{\partial\left[A_j^{(z_j)}\right]}{\partial x} + \left[A_j^{(z_j)}\right]n\frac{-1}{n^2}\frac{\partial n}{\partial x}
$$

$$
= \frac{\partial\left[A_j^{(z_j)}\right]}{\partial x} - \left[A_j^{(z_j)}\right]\frac{1}{n}\frac{\partial n}{\partial x}
$$

$$
= \frac{\partial\left[A_j^{(z_j)}\right]}{\partial x} - \left[A_j^{(z_j)}\right]\frac{1}{-\left[A_j^{(z_j)}\right]+\sqrt{\left[A_j^{(z_j)}\right]^2+4n_i^2}}\left(-1+\frac{\left[A_j^{(z_j)}\right]}{\sqrt{\left[A_j^{(z_j)}\right]^2+4n_i^2}}\right)\frac{\partial\left[A_j^{(z_j)}\right]}{\partial x}
$$

$$= \left[1 + \frac{1}{\sqrt{1 + \left(\frac{4n_i}{\left[A_j^{(z_j)} \right]} \right)^2}} \right] \frac{\partial \left[A_j^{(z_j)} \right]}{\partial x} \tag{45}$$

Therefore, the flux is expressed by a general form independent of whether donor or acceptor as

$$f = -\left(\frac{n}{n_i} \right)^{-z} D_{A_{j-z}} \left(\frac{n}{n_i} \right)^{-z_j} \frac{\partial}{\partial x} \left\{ \left(\frac{n}{n_i} \right)^{z_j} \left[A_j^{(z_j)} \right] \right\}$$

$$= -\left(\frac{n}{n_i} \right)^{-z} D_{A_{j-z}} \left[1 + \frac{1}{\sqrt{1 + \left(\frac{4n_i}{\left[A_j^{(z_j)} \right]} \right)^2}} \right] \frac{\partial \left[A_j^{(z_j)} \right]}{\partial x} \tag{46}$$

The charge of the point defect is thought to be $-2, -1, 0, 1$ [1], and hence the flux is given by

$$f = -\left[1 + \frac{1}{\sqrt{1 + \left(\frac{4n_i}{\left[A_j^{(z_j)} \right]} \right)^2}} \right] \left[D_{A_{j-0}} + \left(\frac{n}{n_i} \right)^{-1} D_{A_{j-1}} + \left(\frac{n}{n_i} \right) D_{A_{j--1}} + \left(\frac{n}{n_i} \right)^2 D_{A_{j--2}} \right] \frac{\partial \left[A_j^{(z_j)} \right]}{\partial x}$$

$$= -\left[1 + \frac{1}{\sqrt{1 + \left(\frac{4n_i}{\left[A_j^{(z_j)} \right]} \right)^2}} \right] \left[D_i^{(x)} + \left(\frac{p}{n_i} \right) D_i^{(p)} + \left(\frac{n}{n_i} \right) D_i^{(m)} + \left(\frac{n}{n_i} \right)^2 D_i^{(mm)} \right] \frac{\partial \left[A_j^{(z_j)} \right]}{\partial x} \tag{47}$$

where

$$D_i^{(x)} \equiv D_{A_j\_0}$$

$$= \frac{1}{2} v_m a^2 r_v \frac{v_1}{v_2} \exp\left[-\frac{\Delta E_m + \Delta E_I + \left(\Delta E_1 - \Delta E_2\right)}{k_B T} \right] = D_i^x \exp\left(-\frac{\Delta E_x}{k_B T} \right) \quad (48)$$

$$D_i^{(p)} \equiv D_{A_j\_1}$$

$$= \frac{1}{2} v_m a^2 r_v \frac{v_1}{v_2} \exp\left[-\frac{\Delta E_m + \Delta E_I + \Delta E_{z=1} + \left(\Delta E_1 - \Delta E_2\right)}{k_B T} \right] = D_i^{(x)} \exp\left(-\frac{\Delta E_p}{k_B T} \right) \quad (49)$$

$$D_i^{(m)} \equiv D_{A_j\_-1}$$

$$= \frac{1}{2} v_m a^2 r_v \frac{v_1}{v_2} \exp\left[-\frac{\Delta E_m + \Delta E_I + \Delta E_{z=-1} + \left(\Delta E_1 - \Delta E_2\right)}{k_B T} \right] = D_i^{(x)} \exp\left(-\frac{\Delta E_m}{k_B T} \right) \quad (50)$$

$$D_i^{(mm)} \equiv D_{A_j\_-2}$$

$$= \frac{1}{2} v_m a^2 r_v \frac{v_1}{v_2} \exp\left[-\frac{\Delta E_m + \Delta E_I + \Delta E_{z=-2} + \left(\Delta E_1 - \Delta E_2\right)}{k_B T} \right] = D_i^{(x)} \exp\left(-\frac{\Delta E_{mm}}{k_B T} \right) \quad (51)$$

From this derivation, the prefactor of the exponential form is the same for any charged diffusion coefficient. However, the shape of the diffusion potential is modified for charged state, and hence $\Delta E_m, \Delta E_1, \Delta E_2, v_m, r_v, v_1, v_2$ may be modified. Therefore, the pre-factor is also different in general and each diffusion coefficient is expressed by the following form as

$$\left.\begin{aligned} D_i^{(x)} &= D_0^{(x)} \exp\left(-\frac{\Delta E_x}{k_B T} \right) \\[2mm] D_i^{(p)} &= D_0^{(p)} \exp\left(-\frac{\Delta E_p}{k_B T} \right) \\[2mm] D_i^{(m)} &= D_0^{(m)} \exp\left(-\frac{\Delta E_m}{k_B T} \right) \\[2mm] D_i^{(mm)} &= D_0^{(mm)} \exp\left(-\frac{\Delta E_{mm}}{k_B T} \right) \end{aligned}\right\} \quad (52)$$

These parameters are determined experimentally.

So far we have considered interstitial Si as a pairing species of the impurity. There is one more point defect: the vacancy. The diffusion associated with vacancies is also considered as shown in Fig. **6**. Although the mechanism is different, the form of the model is the same. Expressing the vacancy and interstitial mechanism, the diffusion coefficient can be expressed by

$$
\left|
\begin{aligned}
D_I &= D_{i\_I}^{(x)} + \left(\frac{p}{n_i}\right) D_{i\_I}^{(p)} + \left(\frac{n}{n_i}\right) D_{i\_I}^{(m)} + \left(\frac{n}{n_i}\right)^2 D_{i\_I}^{(mm)} \\
D_V &= D_{i\_V}^{(x)} + \left(\frac{p}{n_i}\right) D_{i\_V}^{(p)} + \left(\frac{n}{n_i}\right) D_{i\_V}^{(m)} + \left(\frac{n}{n_i}\right)^2 D_{i\_V}^{(mm)}
\end{aligned}
\right.
\tag{53}
$$

The total diffusion coefficient is then expressed by

$$
D = \left(D_{i\_I}^{(x)} + D_{i\_V}^{(x)}\right) + \left(\frac{p}{n_i}\right)\left(D_{i\_I}^{(p)} + D_{i\_V}^{(p)}\right) + \left(\frac{n}{n_i}\right)\left(D_{i\_I}^{(m)} + D_{i\_V}^{(m)}\right) + \left(\frac{n}{n_i}\right)^2 \left(D_{i\_I}^{(mm)} + D_{i\_V}^{(mm)}\right) \tag{54}
$$

If the point defect concentration is at thermal equilibrium, we cannot observe the difference of the point defects, and the summation of them. We then set the summation of them as

$$
D_{i\_I}^{(x)} + D_{i\_V}^{(x)} \equiv D_i^{(x)} \tag{55}
$$

We introduce the parameter f_{ieff} as

$$
D_{i\_I}^{(x)} + D_{i\_V}^{(x)} = f_{Ieff}^{(x)} D_i^{(x)} + \left(1 - f_{Ieff}^{(x)}\right) D_i^{(x)} \tag{56}
$$

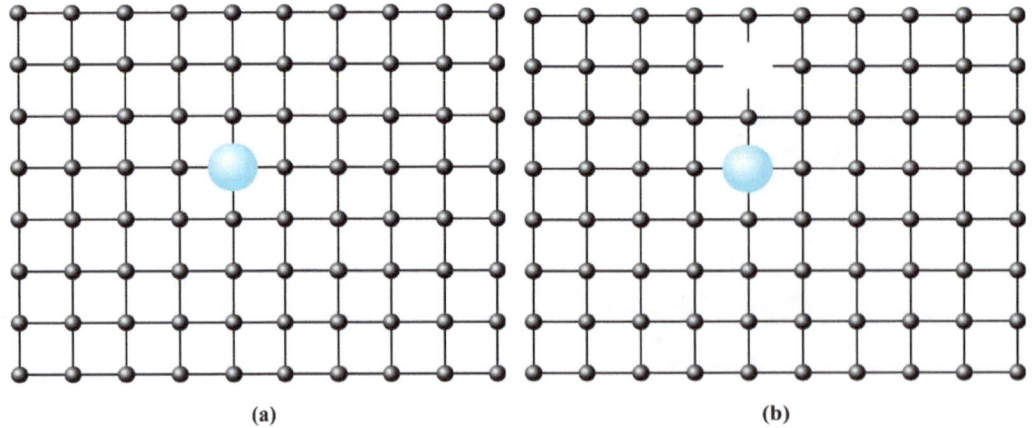

(a)   (b)

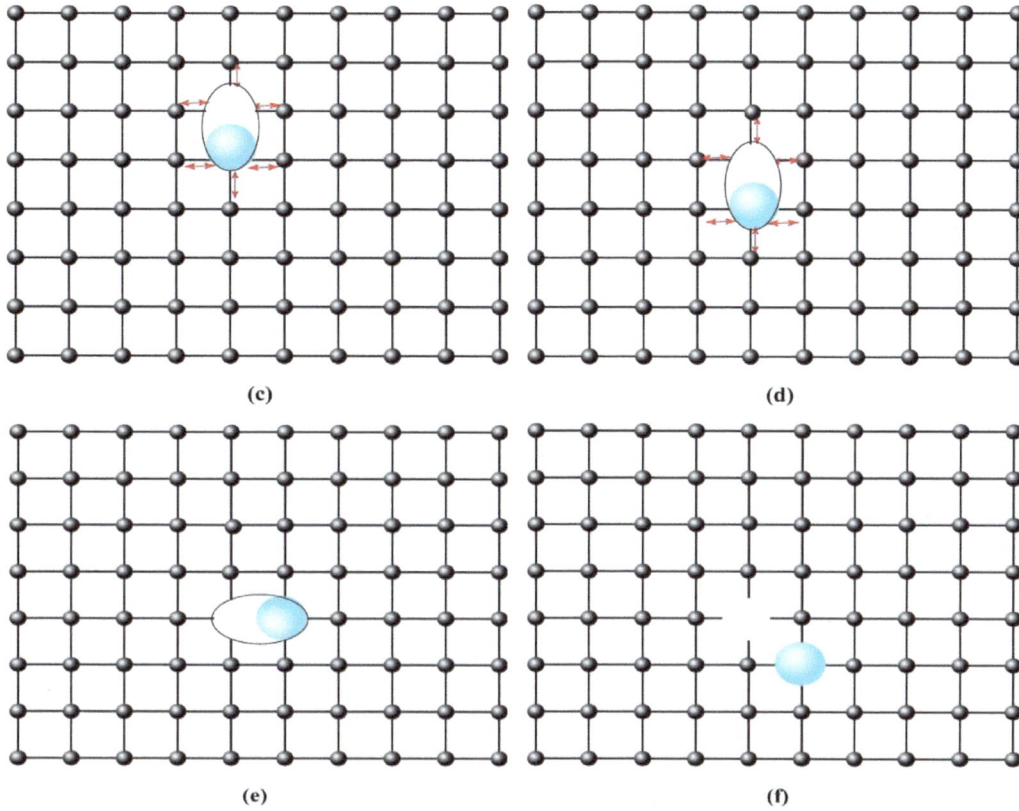

Figure 6: Diffusion animation associated with vacancy paring mechanism. (a) Impurity stays at the lattice site. (b) Diffused vacancy is close to a lattice site impurity by chance. (c) The vacancy and impurity form a pair, which have a chance to move to four directions. (d) The impurity exchanges the pair vacancy and the pair moves down. (e) The impurity exchanges the pair vacancy and moves toright. (f) The pairdissolves and impurity set at the lattice site releasing avacancy.

$f_{Ieff}^{(x)}$ expresses the fraction of the interstitial component of the diffusion coefficient. We apply the coefficient to the other charged components as

$$
\begin{aligned}
D = & \left\lfloor f_{Ieff}^{(x)} D_i^{(x)} + \left(1 - f_{Ieff}^{(x)}\right) D_i^{(x)} \right\rfloor \\
& + \left(\frac{p}{n_i}\right) \left[f_{Ieff}^{(p)} D_i^{(p)} + \left(1 - f_{Ieff}^{(p)}\right) D_i^{(p)} \right] \\
& + \left(\frac{n}{n_i}\right) \left[f_{Ieff}^{(m)} D_i^{(m)} + \left(1 - f_{Ieff}^{(m)}\right) D_i^{(m)} \right] \\
& + \left(\frac{n}{n_i}\right)^2 \left[f_{Ieff}^{(mm)} D_i^{(mm)} + \left(1 - f_{Ieff}^{(mm)}\right) D_i^{(mm)} \right]
\end{aligned}
\tag{57}
$$

This is only rearranging of Eq. 54. However, it does play a role for the diffusion under non thermal equilibrium with respect to point defect concentrations.

Equation 57 implicitly assumes that the thermal equilibrium associated with point defect concentrations is established. However, the concentrations deviate significantly when ion implantation or oxidation processes are performed. We denote the concentrations in thermal equilibrium with as symbol *, and simply assume that the pair concentration is proportional to the point defect concentration, and express the diffusion coefficient under non-thermal equilibrium as

$$
\begin{aligned}
D = &\left| \frac{\left[I^{(0)}\right]}{\left[I^{(0)}\right]^*} f_{Ieff}^{(x)} D_i^{(x)} + \frac{\left[V^{(0)}\right]}{\left[V^{(0)}\right]^*} \left(1 - f_{Ieff}^{(x)}\right) D_i^{(x)} \right| \\
&+ \left(\frac{p}{n_i}\right) \left[\frac{\left[I^{(p)}\right]}{\left[I^{(p)}\right]^*} f_{Ieff}^{(p)} D_i^{(p)} + \frac{\left[V^{(p)}\right]}{\left[V^{(p)}\right]^*} \left(1 - f_{Ieff}^{(p)}\right) D_i^{(p)} \right] \\
&+ \left(\frac{n}{n_i}\right) \left[\frac{\left[I^{(m)}\right]}{\left[I^{(m)}\right]^*} f_{Ieff}^{(m)} D_i^{(m)} + \frac{\left[V^{(m)}\right]}{\left[V^{(m)}\right]^*} \left(1 - f_{Ieff}^{(m)}\right) D_i^{(m)} \right] \\
&+ \left(\frac{n}{n_i}\right)^2 \left[\frac{\left[I^{(mm)}\right]}{\left[I^{(mm)}\right]^*} f_{Ieff}^{(mm)} D_i^{(mm)} + \frac{\left[V^{(mm)}\right]}{\left[V^{(mm)}\right]^*} \left(1 - f_{Ieff}^{(mm)}\right) D_i^{(m)} \right]
\end{aligned}
\tag{58}
$$

It can also be expressed by

$$
\begin{aligned}
D = &\left| \left[\frac{\left[I^{(0)}\right]}{\left[I^{(0)}\right]^*} f_{Ieff}^{(x)} + \frac{\left[V^{(0)}\right]}{\left[V^{(0)}\right]^*} \left(1 - f_{Ieff}^{(x)}\right) \right] D_i^{(x)} \right| \\
&+ \left(\frac{p}{n_i}\right) \left[\frac{\left[I^{(p)}\right]}{\left[I^{(p)}\right]^*} f_{Ieff}^{(p)} + \frac{\left[V^{(p)}\right]}{\left[V^{(p)}\right]^*} \left(1 - f_{Ieff}^{(p)}\right) \right] D_i^{(p)} \\
&+ \left(\frac{n}{n_i}\right) \left[\frac{\left[I^{(m)}\right]}{\left[I^{(m)}\right]^*} f_{Ieff}^{(m)} + \frac{\left[V^{(m)}\right]}{\left[V^{(m)}\right]^*} \left(1 - f_{Ieff}^{(m)}\right) \right] D_i^{(m)} \\
&+ \left(\frac{n}{n_i}\right)^2 \left[\frac{\left[I^{(mm)}\right]}{\left[I^{(mm)}\right]^*} f_{Ieff}^{(mm)} + \frac{\left[V^{(mm)}\right]}{\left[V^{(mm)}\right]^*} \left(1 - f_{Ieff}^{(mm)}\right) \right] D_i^{(mm)}
\end{aligned}
\tag{59}
$$

In general, we can only obtain experimental data associated with the total impurity concentration, but not the concentration of dopant-defect pairs. Therefore, it is impractical to have many ambiguous parameters in the theory when we compare experimental data with a theory.

We assume that the charging reaction is established even in non-thermal equilibrium associated with point defect concentration. The exchange of charges is significantly faster than the reaction of species, which should be a good approximation. Therefore, we assume the followings.

$$
\left\{
\begin{aligned}
\frac{\left\lfloor I^{(0)} \right\rfloor}{\left[I^{(0)} \right]^*} &= \frac{\left\lfloor I^{(p)} \right\rfloor}{\left[I^{(p)} \right]^*} = \frac{\left\lfloor I^{(m)} \right\rfloor}{\left[I^{(m)} \right]^*} = \frac{\left\lfloor I^{(mm)} \right\rfloor}{\left[I^{(mm)} \right]^*} \equiv \frac{[I]}{[I]^*} \\[2mm]
\frac{\left[V^{(0)} \right]}{\left[V^{(0)} \right]^*} &= \frac{\left[V^{(p)} \right]}{\left[V^{(p)} \right]^*} = \frac{\left[V^{(m)} \right]}{\left[V^{(m)} \right]^*} = \frac{\left[V^{(mm)} \right]}{\left[V^{(mm)} \right]^*} \equiv \frac{[V]}{[V]^*}
\end{aligned}
\right.
\tag{60}
$$

Further, we assume the following for simplicity

$$
f_{Ieff}^{(x)} = f_{Ieff}^{(p)} = f_{Ieff}^{(m)} = f_{Ieff}^{(mm)} \equiv f_{Ieff}
\tag{61}
$$

although it lacks physical reason and is not valid for all dopants. The diffusion coefficient is then expressed by

$$
D = \left| \left[\frac{[I]}{[I]^*} f_{Ieff}^{(x)} + \frac{[V]}{[V]^*} \left(1 - f_{Ieff}^{(x)} \right) \right] \left[D_i^{(x)} + \left(\frac{p}{n_i} \right) D_i^{(p)} + \left(\frac{n}{n_i} \right) D_i^{(m)} + \left(\frac{n}{n_i} \right)^2 D_i^{(mm)} \right] \right|
\tag{62}
$$

Considering the drift component, the final expression for the flux associated with total impurity concentration is given by

$$
f = -K_{ele} K_{pdef} \left[D_i^{(x)} + \left(\frac{p}{n_i} \right) D_i^{(p)} + \left(\frac{n}{n_i} \right) D_i^{(m)} + \left(\frac{n}{n_i} \right)^2 D_i^{(mm)} \right] \frac{\partial \left\lfloor A_j^{(z_j)} \right\rfloor}{\partial x}
\tag{63}
$$

where

$$K_{ele} \equiv \left| 1 + \frac{1}{\sqrt{1 + \left(\frac{4n_i}{\left[A_j^{(z_j)} \right]} \right)^2}} \right| \qquad (64)$$

$$K_{pdef} \equiv \frac{[I]}{[I]^*} f_{Ieff} + \frac{[V]}{[V]^*} \left(1 - f_{Ieff} \right) \qquad (65)$$

The diffusion coefficient of B, As, P are expressed by the form of

$$D = D_0 \; [cm^2/s] \; \exp\left(-\frac{\Delta E[eV]}{k_B T} \right) \qquad (66)$$

and corresponding parameter values are reported as shown in Table **1** [1].

It is interesting to compare $D_i^{(x)}$ of B and the theoretical one of

$$D_i^{(x)} = \frac{1}{2} \nu_m a^2 r_v \frac{\nu_1}{\nu_2} \exp\left| -\frac{\Delta E_m + \Delta E_I + \left(\Delta E_1 - \Delta E_2 \right)}{k_B T} \right|$$

D_0 can be roughly evaluated as

$$\begin{aligned}
D_0 &= \frac{1}{2} \nu_m a^2 r_v \frac{\nu_1}{\nu_2} \\
&\approx \frac{1}{2} \nu_m a^2 \\
&= \frac{1}{2} \times 1.3 \times 10^{12} \times \left(0.565 \times 10^{-7} \right)^2 \\
&= 0.002 \left[cm^2\big/ s \right]
\end{aligned}$$

which has a similar order of magnitude as the experimental one in the table.

Table 1: Diffusion parameters of B, As, and P

		B	As	P
D_i^x	D_0	0.037	0.066	3.85
	ΔE	3.46	3.44	3.66

D_i^p	D_0	0.76	-----------	-----------
	ΔE	3.46	-----------	-----------
D_i^m	D_0	-----------	22.9	4.44
	ΔE	-----------	4.1	4.0
D_i^{mm}	D_0	-----------	-----------	44.2
	ΔE	-----------	-----------	4.37

Using the values in Fig. **3** [2], we can evaluate the activation energy as

$$\Delta E = \Delta E_m + \Delta E_I + \left(\Delta E_1 - \Delta E_2 \right)$$
$$= 0.3 + \Delta E_I + 0.4$$
$$= \Delta E_I + 0.7 \left[eV \right]$$

The generation energy of interstitial Si is around 3.0 eV, although smaller and higher values have also been reported in literature.

The diffusion coefficient of Sb and In are reported as shown in Table **2** [3, 4].

Table 2: Diffusion parameters of Sb and In

		Sb	In
D_i^x	D_0	1500	1.443
	ΔE	4.80	3.5645
D_i^p	D_0	13	-----------
	ΔE	4.06	-----------
D_i^m	D_0	-----------	-----------
	ΔE	-----------	-----------

These diffusion coefficients are shown in Fig. **7**. Fig. **8** summarizes the intrinsic diffusion coefficients. There are roughly two groups; rather high diffusion group of B, P, and In, and rather low diffusion of As and Sb.

The diffusion profile is significantly different between high concentration and low concentration region. Fig. **9** shows B diffusion profiles at 1000°C evaluated with FabMeister-IM [5]. The ion implantation doses are 1×10^{13} cm^{-2} (a) and 5×10^{15} cm^{-2} (b). The diffusion depth is significantly deeper for high concentration region, and the shape is also different and the profile looks almost like a rectangular box for the high concentration case.

What causes this difference and what does high concentrationmean in diffusion phenomena? The criterion for that is the intrinsic carrier concentration. If the active concentration is much higher than the intrinsic carrier concentration n_i, the factor p/n_i in Eq. 62 becomes much higher than 1, while it is close to 1 when the impurity concentration is smaller than n_i. The intrinsic carrier concentration is given by [6]

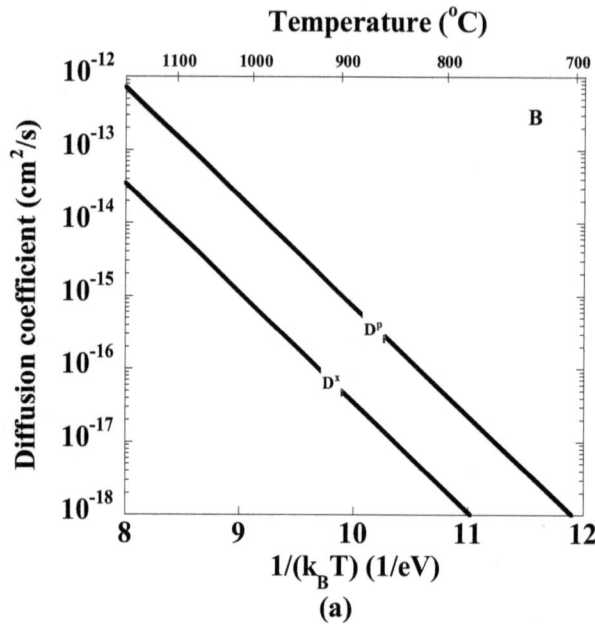

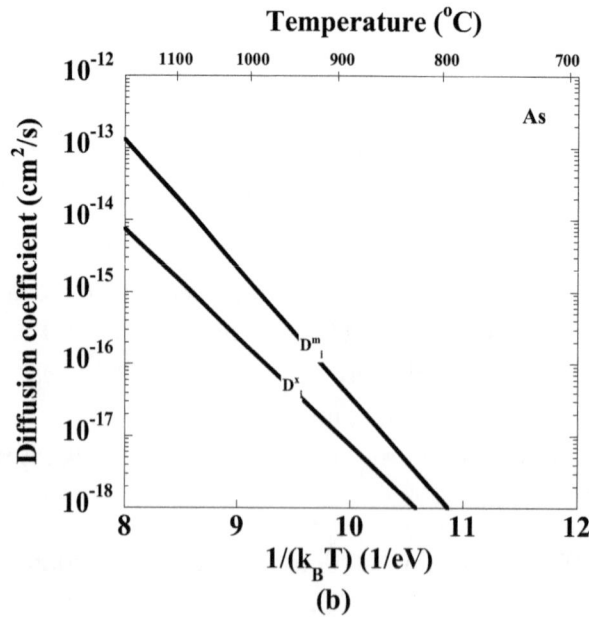

Temperature (^{o}C)

(a)

Temperature (^{o}C)

(b)

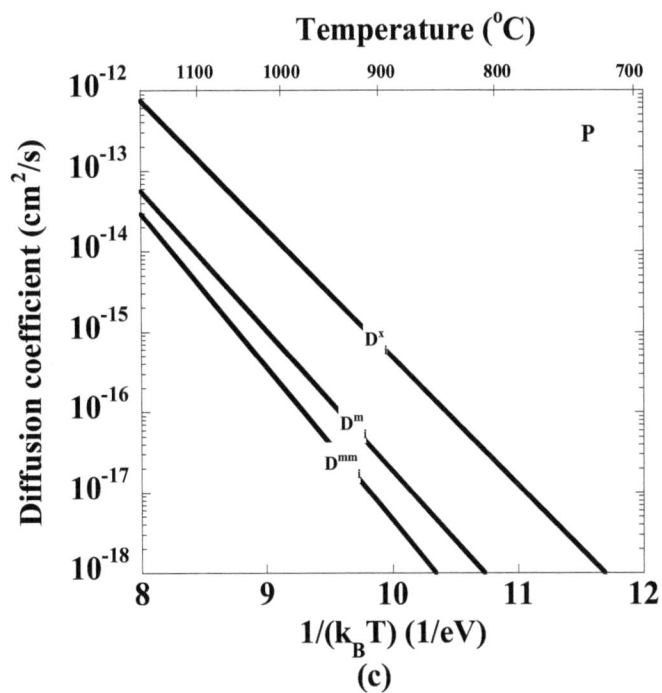

(c)

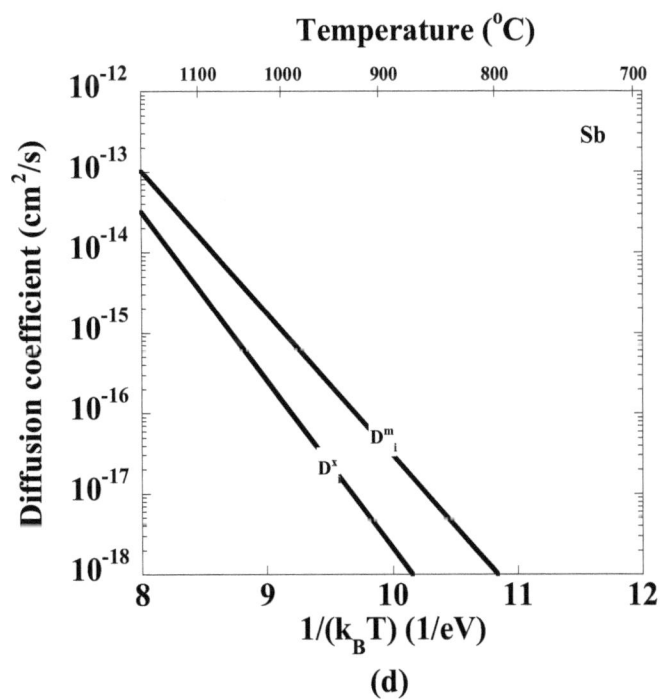

(d)

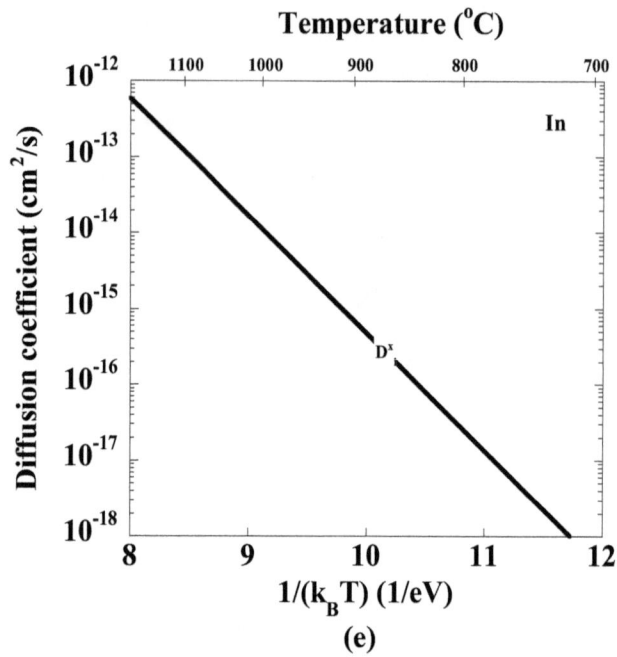

Figure 7: Diffusion coefficients of impurities (a) B, (b) As, (c) P, (d) Sb, (e) In.

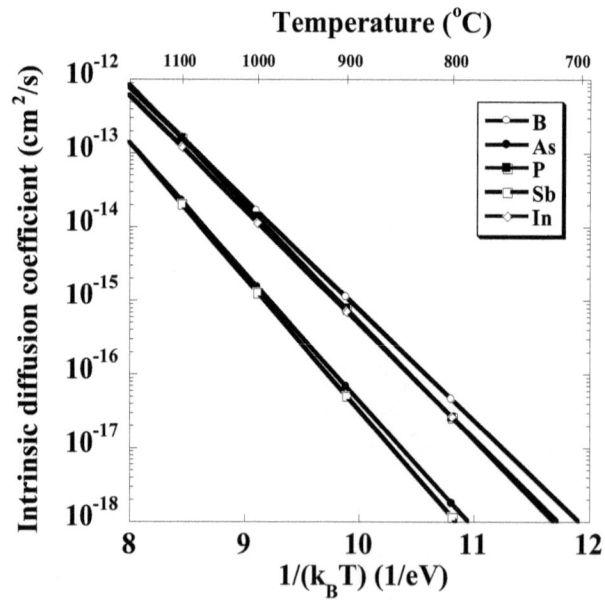

Figure 8: Intrinsic diffusion coefficient in Si substrates of various impurities.

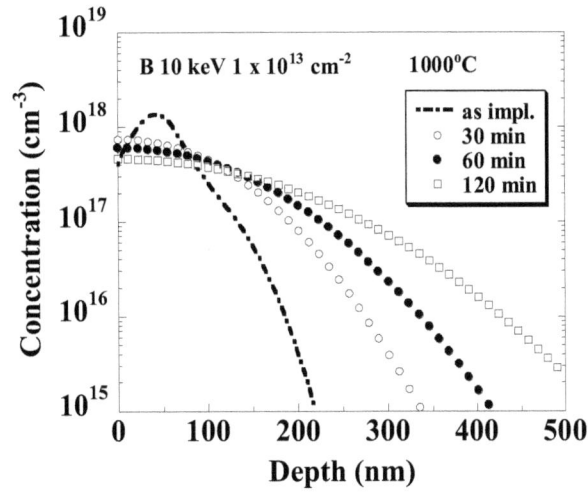

(a)

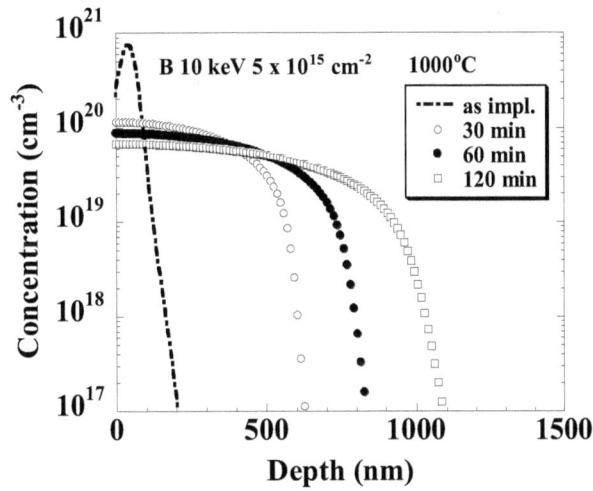

(b)

Figure 9: B diffusion profiles. (a) $1 \times 10^{13} \, cm^{-2}$, (b) $5 \times 10^{15} \, cm^{-2}$.

$$n_i = 3.87 \times 10^{16} T^{\frac{3}{2}} \exp\left[-\frac{0.605\,(eV)}{k_B T} \right] cm^{-3}$$

$$= 2.01 \times 10^{20} \left(\frac{T}{300} \right)^{\frac{3}{2}} \exp\left[-\frac{0.605\,(eV)}{k_B T} \right] cm^{-3}$$

(67)

Fig. **10** shows the dependence of n_i on temperature. It is of the order of $10^{18} cm^{-3}$ in the practical thermal process between 800 and 1000°C. In the low dose case, the diffusion coefficient is constant. In high dose case, the maximum active concentration is of the order of $10^{20} cm^{-3}$, and hence the factor associated high concentration is 100.

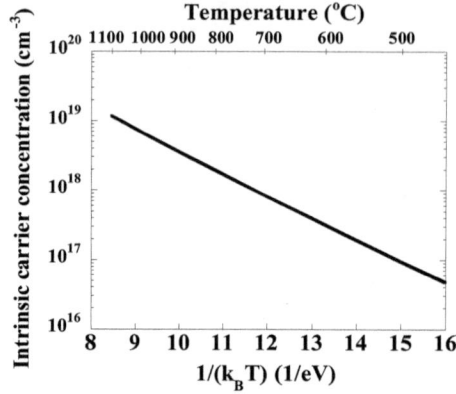

Figure 10: Dependence of intrinsic carrier concentration n_i on temperature.

DERIVATION OF DIFFUSION EQUATION

Once we know a diffusion flux, we can derive a diffusion equation that determines the time evolution of the concentration at each location, and hence redistribution profiles.

The change of the concentration at x is expressed by

$$\frac{\partial N(x)\Delta x}{\partial t} = f(x) - f(x+\Delta x) \tag{68}$$

The diffusion equation is then given by

$$\begin{aligned}\frac{\partial N(x)}{\partial t} &= \frac{f(x)-f(x+\Delta x)}{\Delta x} \\ &= -\frac{\partial f(x)}{\partial x} \\ &= \frac{\partial \left[K_{ele}K_{pdef}D\frac{\partial N(x)}{\partial x}\right]}{\partial x}\end{aligned} \tag{69}$$

If we assume low concentration and thermal equilibrium point defect concentration, K_{ele}, K_{pdef} is 1, and D is constant. The diffusion equation is then simplified as

$$\frac{\partial N(x)}{\partial t} = D\frac{\partial^2 N(x)}{\partial x^2} \tag{70}$$

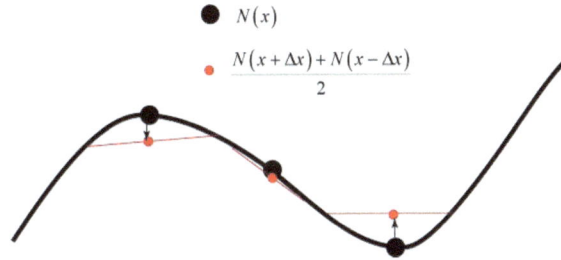

Figure 11: Qualitative approximation of diffusion equation.

We have provided a qualitative image of the diffusion equation. The diffusion equation is expressed by a discrete form of

$$\begin{aligned}
\frac{\partial N(x)}{\partial t} &= D\frac{\partial}{\partial x}\left(\frac{\partial N(x)}{\partial x}\right) \\
&= D\frac{\partial}{\partial x}\left(\frac{\partial N(x)}{\partial x}\bigg|_{x+\Delta x} - \frac{\partial N(x)}{\partial x}\bigg|_{x}\right) \\
&= D\frac{1}{\Delta x}\left(\frac{N(x+\Delta x)-N(x)}{\Delta x} - \frac{N(x)-N(x-\Delta x)}{\Delta x}\right) \\
&= D\frac{2}{(\Delta x)^2}\left(\frac{N(x+\Delta x)+N(x-\Delta x)}{2} - N(x)\right)
\end{aligned} \tag{71}$$

The first term of the left side of Eq. 71 is the average concentration near the $N(x)$. The concentration increases if the concentration is lower than the average concentration and *vice versa* as shown in Fig. **11**.

APPENDIX

There is another diffusion model, whose diffusion potential is described in Fig. **1**. In the model, the diffusion mechanism was proposed for B [7].

When an interstitial Si meets B in the lattice site, B immediately form a BI pair without activation energy. The BI pair does not diffuse. The BI pair then dissolves and interstitial Si occupies the lattice site, and B becomes interstitial B. This interstitial B denoted by B_i diffuses. After a certain step of the diffusion, B forms a BI pair and finally B comes to rest at the lattice site and an interstitial Si is released.

We consider the reaction with z_j charged $A_j^{(z_j)}$ and neutral interstitial Si $I^{(0)}$. We do not consider charged interstitial Si for simplicity.

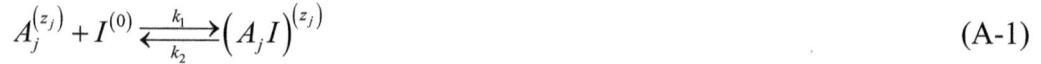

$$A_j^{(z_j)} + I^{(0)} \underset{k_2}{\overset{k_1}{\rightleftarrows}} \left(A_j I\right)^{(z_j)} \tag{A-1}$$

This $\left(A_j I\right)^{(z_j + z)}$ release interstitial B $\left(B_i\right)$. The corresponding reaction is

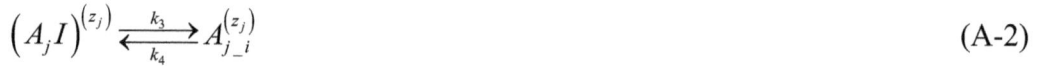

$$\left(A_j I\right)^{(z_j)} \underset{k_4}{\overset{k_3}{\rightleftarrows}} A_{j\_i}^{(z_j)} \tag{A-2}$$

The flux related to the reaction coefficient k_1 is given by

$$\frac{\partial \left[\left(A_j I\right)^{(z_j)}\right]}{\partial t} = \frac{1}{2}\left[A_j^{(z_j)}\right]\frac{\left[I^{(0)}\right]}{\left[Si\right]}v_1 \exp\left(-\frac{\Delta E_1}{k_B T}\right) \tag{A-3}$$

The flux related to the reaction coefficient k_2 is given by

$$\frac{\partial \left[\left(A_j I\right)^{(z_j)}\right]}{\partial t} = -\frac{1}{2}\left[\left(A_j I\right)^{(z_j)}\right]v_2 \exp\left(-\frac{\Delta E_2}{k_B T}\right) \tag{A-4}$$

The flux related to the reaction coefficient k_3 is given by

$$\frac{\partial \left[\left(A_j I\right)^{(z_j)}\right]}{\partial t} = -\frac{1}{2}\left[\left(A_j I\right)^{(z_j)}\right]v_2 \exp\left(-\frac{\Delta E_3}{k_B T}\right) \tag{A-5}$$

The flux related to the reaction coefficient k_4 is given by

$$\frac{\partial \left[\left(A_j I \right)^{(z_j)} \right]}{\partial t} = \frac{1}{2} \left[\left(A_{j\_i} \right)^{(z_j)} \right] v_3 \exp\left(-\frac{\Delta E_4}{k_B T} \right) \tag{A-6}$$

Assuming the reaction associated with $\left(A_{j\_i} \right)^{(z_j)}$ is in thermal equilibrium, we obtain

$$\left[\left(A_{j\_i} \right)^{(z_j)} \right] = \left[\left(A_j I \right)^{(z_j)} \right] \frac{v_2}{v_3} \exp\left(-\frac{\Delta E_3 - \Delta E_4}{k_B T} \right) \tag{A-7}$$

Assuming the reaction associated with $\left(A_j I \right)^{(z_j)}$ is in thermal equilibrium, we obtain form k_1, k_2 as

$$\left[\left(A_j I \right)^{(z_j)} \right] = \left[A_j^{(z_j)} \right] \frac{\left[I^{(0)} \right]}{[Si]} \frac{v_1}{v_2} \exp\left(-\frac{\Delta E_1 - \Delta E_2}{k_B T} \right) \tag{A-8}$$

Therefore, the diffusion species $\left(A_{j\_i} \right)^{(z_j)}$ is related to the observed $A_j^{(z_j)}$ as

$$\begin{aligned}
\left[\left(A_{j\_i} \right)^{(z_j)} \right] &= \left[A_j^{(z_j)} \right] \frac{\left[I^{(0)} \right]}{[Si]} \frac{v_1}{v_3} \exp\left(-\frac{\left(\Delta E_1 - \Delta E_2 \right) + \left(\Delta E_3 - \Delta E_4 \right)}{k_B T} \right) \\
&= \left[A_j^{(z_j)} \right] \frac{v_1}{v_3} \exp\left(-\frac{\Delta E_I + \left(\Delta E_1 - \Delta E_2 \right) + \left(\Delta E_3 - \Delta E_4 \right)}{k_B T} \right)
\end{aligned} \tag{A-9}$$

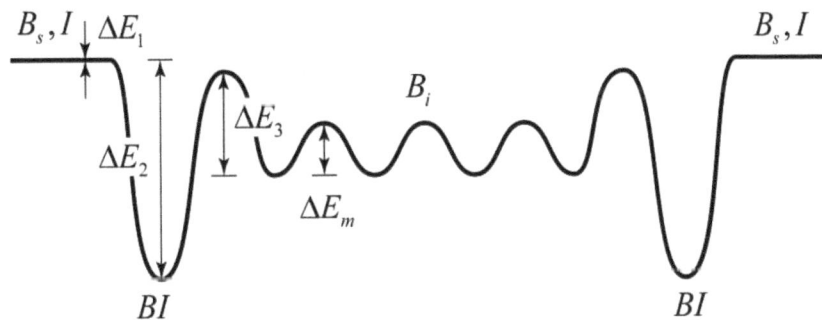

Figure A-1: Diffusion potential for B_i diffusion potential.

The diffusion flux of $\left(A_{j\_i} \right)^{(z_j)}$ is then given by

$$f = -\frac{1}{2}v_m a^2 \exp\left[-\frac{\Delta E_m}{k_B T}\right]\frac{\partial\left[\left(A_{j\_i}\right)^{(z_j)}\right]}{\partial x}$$

$$= -\frac{1}{2}v_m a^2 \exp\left[-\frac{\Delta E_m}{k_B T}\right]\frac{\partial\left[A_j^{(z_j)}\right]\frac{v_1}{v_3}\exp\left(-\frac{\Delta E_I + \left(\Delta E_1 - \Delta E_2\right) + \left(\Delta E_3 - \Delta E_4\right)}{k_B T}\right)}{\partial x} \quad \text{(A-10)}$$

$$= -\frac{1}{2}v_m \frac{v_1}{v_3}a^2 \exp\left(-\frac{\Delta E_m + \Delta E_I + \left(\Delta E_1 - \Delta E_2\right) + \left(\Delta E_3 - \Delta E_4\right)}{k_B T}\right)\frac{\partial\left[A_j^{(z_j)}\right]}{\partial x}$$

The physical meaning of preexponential factor and activation energy is different depending on the mechanism, but the form is the same.

REFERENCES

[1]   Sorab K. Ghandhi, VLSI Fabrication Principles, John Wiley & Sons, New York, p. 247, 1994.

[2]   Jing Zhu, Tomas Diaz dela Rubia, L. H. Yang, and Christian Mailhiot, "Ab initio pseudopotential calculations of B diffusion and pairing in Si," PHYSICAL REVIEW B VOLUME 54, NUMBER 7, pp. 4741-4755, 1996.

[3]   K. Suzuki, Hiroko Tashiro, and Takayuki Aoyama," Diffusion coefficient of indium in Si substrates and analytical redistribution model," Solid-State Electronics、 vol.43, pp. 27-31, 1999

[4]   K. Suzuki, Hiroko Tashiro, and Takayuki Aoyama," Sb diffusion in heavily doped Si substrates," J. Electrochem. Society, vol. 146, pp. 336-338, 1999.

[5]   Mizuho Information & Research Institute, Inc.: Overview of the FabMeister-IM ion implantation profile simulator.
      http://www.mizuho-ir.co.jp/solution/research/semiconductor/fabmeister/ion/index.html

[6]   F. J. Morin and J. P. Maita, "Electrical properties of silicon containing arsenic and boron," Physical Review, Vol. 96, pp. 28-35, 1954.

[7]   W. Windl, M. M. Bunea,R. Stumpf,S. T. Dunham, and M. P. Masquelier, "First-Principles Study of Boron Diffusion in Silicon," PHYSICAL REVIEW LETTERS, VOLUME 83, NUMBER 21, pp. 4345-4348, 1999.

Send Orders for Reprints to reprints@benthamscience.net

Ion Implantation and Activation, Vol. 3, 2013, 35-55

CHAPTER 2

Paring Diffusion Equation

Abstract: In this section, we consider reaction between impurity, point defects, and charges in diffusion phenomenon and diffusion equations for both of impurity and point defects. There are many levels of sophistication of treatment of the interaction of impurity and point defects. We perform approximations step by step, and derive diffusion models in various supplication levels.

Keywords: Ion implantation, diffusion, transient enhanced diffusion, pairing, point defects, diffusion flux, diffusivity, thermal equilibrium, electric field, mobility, Boltzmann constant, Einstein relationship, electron, hole, interstitial Si, vacancy, balance equation, five stream model, three stream model.

INTRODUCTION

Point defects exist in substrate, and they play a role of vehicle of impurities by paring each other. Ion implantation induces significant excess point defects, and hence they significantly influence diffusion of impurities. On the other hand, the diffusivity of point defects is much higher than the impurity diffusion coefficient [1] although the value is not well established and scattered in many orders. The excess point defects vanish in the early stage of the impurity diffusion. Therefore, we observe significant high or low diffusion in this quite short time region, and then observe normal diffusion under thermal equilibrium point defect concentration after certain time period. This early stage of diffusion under non-thermal equilibrium point defect concentrations is called as transient diffusion, and related diffusion equations are developed [2]. We should trace impurity point defects pairs and point defects themselves simultaneously to grasp the transient diffusion phenomenon.

DIFFUSION FLUX

Flux of species J is given by

$$J = -D\nabla N + z\varepsilon\mu N \qquad (1)$$

where D is the diffusion coefficient, z is the charging number, ε is the electric field, μ is the mobility, N is the concentration of diffusion species.

Kunihiro Suzuki

The first term in Eq. 1 is the diffusion component and the second drift one. Electron concentration n is related to potential ϕ as

$$n = n_i \exp\left(\frac{q\phi}{k_B T}\right) \tag{2}$$

where q is the charge of electron, k_B is the Boltzmann constant, and T is the absolute temperature. The electric field ε is then given by

$$\varepsilon = -\nabla\phi = -\frac{k_B T}{q}\nabla\ln\left(\frac{n}{n_i}\right) \tag{3}$$

Using Einstein relationship

$$D = \frac{k_B T}{q}\mu \tag{4}$$

we obtain a form of the flux as

$$
\begin{aligned}
J &= -D\nabla N - \frac{k_B T}{q}\mu N\nabla\ln\left(\frac{n}{n_i}\right)^z \\
&= -D\nabla N - DN\nabla\ln\left(\frac{n}{n_i}\right)^z \\
&= -D\left[\nabla N + DN\left(\frac{n}{n_i}\right)^{-z}\nabla\left(\frac{n}{n_i}\right)^z\right] \\
&= -D\left(\frac{n}{n_i}\right)^{-z}\left[\left(\frac{n}{n_i}\right)^z\nabla N + DN\nabla\left(\frac{n}{n_i}\right)^z\right] \\
&= -D\left(\frac{n}{n_i}\right)^{-z}\nabla\left[N\left(\frac{n}{n_i}\right)^z\right]
\end{aligned}
\tag{5}
$$

REACTION

We describe reactions between impurities, point defects, and electron and holes. The same symbol is used for a species and the corresponding concentration of the species.

Paring between impurity and point defects

z_j-changed impurity $A_j^{[z_j]}$ reacts with z-charged interstitial Si $I^{[z]}$ and form $(z_j + y)$-charged impurity interstitial Si pair of $(A_j I)^{[z_j + y]}$, which is expressed by

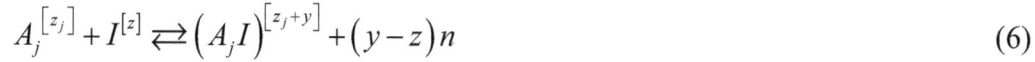

$$A_j^{[z_j]} + I^{[z]} \rightleftarrows (A_j I)^{[z_j + y]} + (y - z)n \tag{6}$$

We denote this reaction rate as R_{1jzy}, which is given by

$$R_{1jzy} = h_{1jzy}\left[A_j^{[z_j]} I^{[z]} - k_{1jzy}(A_j I)^{[z_j + y]}\left(\frac{n}{n_i}\right)^{(y-z)} \right] \tag{7}$$

h_{1jzy} is the transport coefficient, and k_{1jzy} is the thermal equilibrium constant associated with the reaction. We use similar notations for the other reactions here after.

Similarly, z_j-changed impurity $A_j^{[z_j]}$ reacts with z-charged vacancy $V^{[z]}$ and form $(z_j + y)$ charged impurity vacancy pair of $(A_j V)^{[z_j + y]}$, which is expressed by

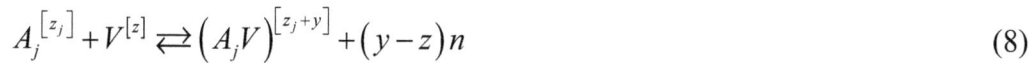

$$A_j^{[z_j]} + V^{[z]} \rightleftarrows (A_j V)^{[z_j + y]} + (y - z)n \tag{8}$$

We denote this reaction rate as R_{2jzy}, which is given by

$$R_{2jzy} = h_{2jzy}\left[A_j^{[z_j]} V^{[z]} - k_{2jzy}(A_j V)^{[z_j + y]}\left(\frac{n}{n_i}\right)^{(y-z)} \right] \tag{9}$$

IONIZATION OF POINT DEFECTS AND IMPURITY POINT DEFECTS PAIR

We consider ionization of point defects and point defect impurity pair.

$(z_j + z)$-changed $(A_j I)^{[z_j + z]}$ is charged and form a $(z_j + y)$-charged pair of $(A_j I)^{[z_j + y]}$, which is expressed by

$$(A_j I)^{[z_j + z]} \rightleftarrows (A_j I)^{[z_j + y]} + (y - z)n \tag{10}$$

We denote this reaction rate as R_{3jzy}, which is given by

$$R_{3jzy} = h_{3jzy} \left[(A_j V)^{[z_j + z]} - k_{3jzy} (A_j V)^{[z_j + y]} \left(\frac{n}{n_i} \right)^{(y-z)} \right] \tag{11}$$

Similarly, the $(z_j + z)$-charged $(A_j V)^{[z_j + z]}$ is charged and form a $(z_j + y)$-charged pair of $(A_j V)^{[z_j + y]}$, which is expressed by

$$(A_j V)^{[z_j + z]} \rightleftarrows (A_j V)^{[z_j + y]} + (y - z)n \tag{12}$$

We denote this reaction rate as R_{4jzy}, which is given by

$$R_{4jzy} = h_{4jzy} \left[(A_j V)^{[z_j + z]} - k_{4jzy} (A_j V)^{[z_j + y]} \left(\frac{n}{n_i} \right)^{(y-z)} \right] \tag{13}$$

We then consider the ionization of point defects. z charged $I^{[z]}$ is charged and form y charged $I^{[y]}$, which is expressed by

$$I^{[z]} \rightleftarrows I^{[y]} + (y - z)n \tag{14}$$

We denote this reaction rate as R_{5jzy}, which is given by

$$R_{5zy} = h_{5zy} \left[I^{[z]} - k_{5zy} I^{[y]} \left(\frac{n}{n_i} \right)^{y-z} \right] \tag{15}$$

Similarly, z charged $V^{[z]}$ is charged and form y charged $V^{[y]}$, which is expressed by

$$V^{[z]} \rightleftharpoons V^{[y]} + (y - z)n \tag{16}$$

We denote this reaction rate as R_{6jzy}, which is given by

$$R_{6zy} = h_{6zy}\left[V^{[z]} - k_{6zy}V^{[y]}\left(\frac{n}{n_i}\right)^{y-z} \right] \tag{17}$$

RECOMBINATION

We consider recombination between I and V, $A_j I$ and V, $A_j V$ and I. $A_j I$ and $A_j V$ concentrations are quite low in general, and we neglect the recombination of them.

The recombination between I and V is described as

$$I^{[z]} + V^{[y]} \rightleftharpoons -(y + z)n \tag{18}$$

We denote this reaction rate as R_{7zy}, which is given by

$$R_{7zy} = h_{7zy}\left[I^{[z]}V^{[y]} - k_{7zy}\left(\frac{n}{n_i}\right)^{-(y+z)} \right] \tag{19}$$

The recombination between $A_j I$ and V is described as

$$\left(A_j I\right)^{[z_j + z]} + V^{[y]} \rightleftharpoons A_j^{[z_j]} - (y + z)n \tag{20}$$

We denote this reaction rate as R_{8jzy}, which is given by

$$R_{8jzy} = h_{8jzy}\left[\left(A_j I\right)^{[z_j + z]}V^{[y]} - k_{8jzy}A_j^{[z_j]}\left(\frac{n}{n_i}\right)^{-(y+z)} \right] \tag{21}$$

The recombination between A_jV and I is described as

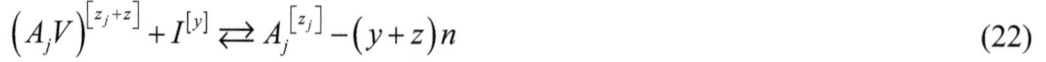

$$\left(A_jV\right)^{[z_j+z]} + I^{[y]} \rightleftarrows A_j^{[z_j]} - (y+z)n \tag{22}$$

We denote this reaction rate as $R_{9\,jzy}$, which is given by

$$R_{9\,jzy} = h_{9\,jzy}\left[\left(A_jV\right)^{[z_j+z]} I^{[y]} - k_{9\,jzy} A_j^{[z_j]}\left(\frac{n}{n_i}\right)^{-(y+z)}\right] \tag{23}$$

BALANCE EQUATIONS

The balance equations for each species are then given by

$I^{[z]}$:

$$\frac{\partial I^{[z]}}{\partial t} + div\sum_z J_{I^{[z]}} = -\sum_{j_y} R_{1\,jzy} - \sum_y R_{5zy} + \sum_y R_{5yz} - \sum_y R_{7zy} - \sum_y R_{9\,jyz} \tag{24}$$

where

$$J_{I^{[z]}} = -D_{I^{[z]}}\left(\frac{n}{n_i}\right)^{-z} \nabla\left[I^{[z]}\left(\frac{n}{n_i}\right)^z\right] \tag{25}$$

$V^{[z]}$:

$$\frac{\partial V^{[z]}}{\partial t} + div\sum_z J_{V^{[z]}} = -\sum_y R_{2\,jzy} - \sum_y R_{6zy} + \sum_y R_{6yz} - \sum_y R_{7yz} - \sum_y R_{8\,jyz} \tag{26}$$

where

$$J_{V^{[z]}} = -D_{V^{[z]}}\left(\frac{n}{n_i}\right)^{-z} \nabla\left[V^{[z]}\left(\frac{n}{n_i}\right)^z\right] \tag{27}$$

$\left(A_jI\right)^{[z_j+z]}$:

$$\frac{\partial \left(A_j I\right)^{\left[z_j+z\right]}}{\partial t} + div\sum_z J_{\left(A_j I\right)^{\left[z_j+z\right]}} = \sum_y R_{1\,jyz} - \sum_y R_{3\,jzy} + \sum_y R_{3\,jyz} - \sum_y R_{8\,jzy} \tag{28}$$

where

$$J_{\left(A_j I\right)^{\left[z_j+z\right]}} = -D_{\left(A_j I\right)^{\left[z_j+z\right]}} \left(\frac{n}{n_i}\right)^{-\left(z_j+z\right)} \nabla\left[\left(A_j I\right)^{\left[z_j+z\right]}\left(\frac{n}{n_i}\right)^{\left(z_j+z\right)}\right] \tag{29}$$

$$\left(A_j V\right)^{\left[z_j+z\right]} :$$

$$\frac{\partial \left(A_j V\right)^{\left[z_j+z\right]}}{\partial t} + div\sum_z J_{\left(A_j V\right)^{\left[z_j+z\right]}} = \sum_y R_{2\,jyz} - \sum_y R_{4\,jzy} + \sum_y R_{4\,jyz} - \sum_y R_{9\,jzy} \tag{30}$$

where

$$J_{\left(A_j V\right)^{\left[z_j+z\right]}} = -D_{\left(A_j V\right)^{\left[z_j+z\right]}} \left(\frac{n}{n_i}\right)^{-\left(z_j+z\right)} \nabla\left[\left(A_j V\right)^{\left[z_j+z\right]}\left(\frac{n}{n_i}\right)^{\left(z_j+z\right)}\right] \tag{31}$$

$$A_j^{\left[z_j\right]} :$$

$$\frac{\partial A_j^{\left[z_j\right]}}{\partial t} = -\sum_{yz} R_{1\,jzy} - \sum_{yz} R_{2\,jzy} + \sum_{yz} R_{8\,jzy} + \sum_{yz} R_{9\,jzy} \tag{32}$$

where we regards that the impurity itself does not move, and hence corresponding divergence component does not exist.

THERMAL EQUILIBRIUM IN IONIZATION

Electrons and holes move much faster than impurity and point defects in solid. Therefore, we can approximate that the thermal equilibrium in ionization is established. Therefore, we set

$$R_{3\,jzy} = R_{4\,jzy} = R_{5zy} = R_{6zy} = 0 \tag{33}$$

Let us consider $R_{3jzy} = 0$, that is

$$R_{3jzy} = h_{3jzy}\left[\left(A_jI\right)^{\left[z_j+z\right]} - k_{3jzy}\left(A_jI\right)^{\left[z_j+y\right]}\left(\frac{n}{n_i}\right)^{y-z}\right] = 0 \tag{34}$$

We then obtain

$$\left(A_jI\right)^{\left[z_j+z\right]} = k_{3jzy}\left(A_jI\right)^{\left[z_j+y\right]}\left(\frac{n}{n_i}\right)^{y-z} \tag{35}$$

We express each charged species concentrations with respect to the one with $y = 0$.

$$\left(A_jI\right)^{\left[z_j+z\right]} = k_{3jz0}\left(A_jI^{[0]}\right)^{\left[z_j\right]}\left(\frac{n}{n_i}\right)^{-z} \tag{36}$$

where we denote paring of impurity and neutral interstitial Si as $\left(A_jI^{[0]}\right)^{\left[z_j\right]}$ instead of $\left(A_jI\right)^{\left[z_j\right]}$ to clarify it as it is.

Similarly, $R_{4jzy} = 0, R_{5zy} = 0, R_{6zy} = 0$ and we obtain

$$\left(A_jV\right)^{\left[z_j+z\right]} = k_{4jz0}\left(A_jV^{[0]}\right)^{\left[z_j\right]}\left(\frac{n}{n_i}\right)^{-z} \tag{37}$$

where we denote paring of impurity and neutral vacancy as $\left(A_jV^{[0]}\right)^{\left[z_j\right]}$ instead of $\left(A_jV\right)^{\left[z_j\right]}$ to clarify it as it is.

$$I^{[z]} = k_{5z0}I^{[0]}\left(\frac{n}{n_i}\right)^{-z} \tag{38}$$

$$V^{[z]} = k_{6z0}V^{[0]}\left(\frac{n}{n_i}\right)^{-z} \tag{39}$$

Under these approximations, species with various charged states are not independent, but related to corresponding thermal equilibrium coefficients.

Therefore, we can treat the sum of the various charged species. The corresponding balance equations are given by follows.

$$\frac{\partial}{\partial t}\sum_z I^{[z]} + div\sum_z J_{I^{[z]}} = -\sum_{jzy} R_{1\,jzy} - \sum_{zy} R_{7\,zy} - \sum_{jzy} R_{9\,jzy} \tag{40}$$

where

$$\sum_z J_{I^{[z]}} = -\sum_z D_{I^{[z]}}\left(\frac{n}{n_i}\right)^{-z} \nabla\left[I^{[z]}\left(\frac{n}{n_i}\right)^z\right]$$

$$= -\sum_z D_{I^{[z]}}\left(\frac{n}{n_i}\right)^{-z} \nabla\left[k_{5z0}I^{[0]}\left(\frac{n}{n_i}\right)^{-z}\left(\frac{n}{n_i}\right)^z\right]$$

$$= -\sum_z k_{5z0}D_{I^{[z]}}\left(\frac{n}{n_i}\right)^{-z} \nabla I^{[0]} \tag{41}$$

$$\frac{\partial}{\partial t}\sum_z V^{[z]} + div\sum_z J_{V^{[z]}} = -\sum_{jzy} R_{2\,jzy} - \sum_{zy} R_{7\,yz} - \sum_{jzy} R_{8\,jyz} \tag{42}$$

where

$$\sum_z J_{V^{[z]}} = -\sum_z D_{V^{[z]}}\left(\frac{n}{n_i}\right)^{-z} \nabla\left[V^{[z]}\left(\frac{n}{n_i}\right)^z\right]$$

$$= -\sum_z D_{V^{[z]}}\left(\frac{n}{n_i}\right)^{-z} \nabla\left[k_{6z0}V^{[0]}\left(\frac{n}{n_i}\right)^{-z}\left(\frac{n}{n_i}\right)^z\right]$$

$$= -\sum_z k_{6z0}D_{V^{[z]}}\left(\frac{n}{n_i}\right)^{-z} \nabla V^{[0]} \tag{43}$$

$$\frac{\partial}{\partial t}\sum_z \left(A_j I\right)^{[z_j+z]} + div\sum_z J_{\left(A_j I\right)^{[z_j+z]}} = -\sum_{zy} R_{1\,jyz} - \sum_{zy} R_{8\,jzy} \tag{44}$$

where

$$
\sum_z J_{(A_j I)^{[z_j+z]}} = -\sum_z D_{(A_j I)^{[z_j+z]}} \left(\frac{n}{n_i}\right)^{-(z_j+z)} \nabla\left[\left(A_j I\right)^{[z_j+z]}\left(\frac{n}{n_i}\right)^{(z_j+z)}\right]
$$

$$
= -\sum_z D_{(A_j I)^{[z_j+z]}} \left(\frac{n}{n_i}\right)^{-(z_j+z)} \nabla\left[k_{3jz0}\left(A_j I^{(0)}\right)^{[z_j]}\left(\frac{n}{n_i}\right)^{-z}\left(\frac{n}{n_i}\right)^{(z_j+z)}\right]
$$

$$
= -\sum_z k_{3jz0} D_{(A_j I)^{[z_j+z]}} \left(\frac{n}{n_i}\right)^{-(z_j+z)} \nabla\left[\left(A_j I^{(0)}\right)^{[z_j]}\left(\frac{n}{n_i}\right)^{(z_j)}\right]
$$

$$
= -\sum_z k_{3jz0} D_{(A_j I)^{[z_j+z]}} \left(\frac{n}{n_i}\right)^{-(z_j+z)} \left\{\left(\frac{n}{n_i}\right)^{z_j} \nabla\left(A_j I^{[0]}\right)^{[z_j]} + \left(A_j I^{[0]}\right)^{[z_j]} \nabla\left(\frac{n}{n_i}\right)^{z_j}\right\}
$$

$$
= -\sum_z k_{3jz0} D_{(A_j I)^{[z_j+z]}} \left(\frac{n}{n_i}\right)^{-z} \left\{\nabla\left(A_j I^{[0]}\right)^{[z_j]} + \left(A_j I^{[0]}\right)^{[z_j]}\left(\frac{n}{n_i}\right)^{-z_j} \nabla\left(\frac{n}{n_i}\right)^{z_j}\right\}
$$

$$
= -\sum_z k_{3jz0} D_{(A_j I)^{[z_j+z]}} \left(\frac{n}{n_i}\right)^{-z} \left\{\nabla\left(A_j I^{(0)}\right)^{[z_j]} + z_j\left(A_j I^{[0]}\right)^{[z_j]} \nabla\left[\ln\left(\frac{n}{n_i}\right)\right]\right\} \tag{45}
$$

$$
\frac{\partial}{\partial t}\sum_z \left(A_j V\right)^{[z_j+z]} + div\sum_z J_{(A_j V)^{[z_j+z]}} = \sum_{zy} R_{2jyz} - \sum_{zy} R_{9jzy} \tag{46}
$$

where

$$\sum_z J_{(A_jV)^{[z_j+z]}}$$

$$= -\sum_z D_{(A_jV)^{[z_j+z]}} \left(\frac{n}{n_i}\right)^{-(z_j+z)} \nabla\left[(A_jV)^{[z_j+z]}\left(\frac{n}{n_i}\right)^{(z_j+z)}\right]$$

$$= -\sum_z D_{(A_jV)^{[z_j+z]}} \left(\frac{n}{n_i}\right)^{-(z_j+z)} \nabla\left[k_{4jz0}\left(A_jV^{[0]}\right)^{[z_j]}\left(\frac{n}{n_i}\right)^{-z}\left(\frac{n}{n_i}\right)^{(z_j+z)}\right]$$

$$= -\sum_z k_{4jz0} D_{(A_jV)^{[z_j+z]}} \left(\frac{n}{n_i}\right)^{-(z_j+z)} \nabla\left[\left(A_jV^{[0]}\right)^{[z_j]}\left(\frac{n}{n_i}\right)^{(z_j)}\right]$$

$$= -\sum_z k_{4jz0} D_{(A_jV)^{[z_j+z]}} \left(\frac{n}{n_i}\right)^{-(z_j+z)} \left\{\left(\frac{n}{n_i}\right)^{z_j}\nabla\left(A_jV^{[0]}\right)^{[z_j]}+\left(A_jV^{[0]}\right)^{[z_j]}\nabla\left(\frac{n}{n_i}\right)^{z_j}\right\}$$

$$= -\sum_z k_{4jz0} D_{(A_jV)^{[z_j+z]}} \left(\frac{n}{n_i}\right)^{-z} \left\{\nabla\left(A_jV^{[0]}\right)^{[z_j]}+\left(A_jV^{[0]}\right)^{[z_j]}\left(\frac{n}{n_i}\right)^{-z_j}\nabla\left(\frac{n}{n_i}\right)^{z_j}\right\}$$

$$= -\sum_z k_{4jz0} D_{(A_jV)^{[z_j+z]}} \left(\frac{n}{n_i}\right)^{-z} \left\{\nabla\left(A_jV^{[0]}\right)^{[z_j]}+z_j\left(A_jV^{[0]}\right)^{[z_j]}\nabla\left[\ln\left(\frac{n}{n_i}\right)\right]\right\} \tag{47}$$

$$\frac{\partial A_j^{[z_j]}}{\partial t} = -\sum_{zy} R_{1jzy} - \sum_{yz} R_{2jzy} + \sum_{zy} R_{8jzy} + \sum_{zy} R_{9jzy} \tag{48}$$

We should solve coupled Eqs. of 40, 42, 44, 46, and 48, which is called as 5 stream model.

Under these approximations, the recombination rate are reduced to

$$R_{1jzy} = h_{1jzy}\left[A_j^{[z_j]}I^{[z]} - k_{1jzy}\left(A_jI\right)^{[z_j+y]}\left(\frac{n}{n_i}\right)^{y-z}\right]$$

$$= h_{1jzy}\left[A_j^{[z_j]}k_{5z0}I^{[0]}\left(\frac{n}{n_i}\right)^{-z} - k_{1jzy}k_{3jz0}\left(A_jI^{[0]}\right)^{[z_j]}\left(\frac{n}{n_i}\right)^{-y}\left(\frac{n}{n_i}\right)^{y-z}\right] \tag{49}$$

$$= h_{1jzy}k_{5z0}\left(\frac{n}{n_i}\right)^{-z}\left[A_j^{[z_j]}I^{[0]} - \frac{k_{1jzy}k_{3jz0}}{k_{5z0}}\left(A_jI^{[0]}\right)^{[z_j]}\right]$$

$$
\begin{aligned}
R_{2jzy} &= h_{2jzy}\left[A_j^{[z_j]}V^{[z]} - k_{2jzy}\left(A_jV\right)^{[z_j+y]}\left(\frac{n}{n_i}\right)^{y-z}\right] \\
&= h_{2jzy}\left[A_j^{[z_j]}k_{6z0}V^{[0]}\left(\frac{n}{n_i}\right)^{-z} - k_{2jzy}k_{4jz0}\left(A_jV^{[0]}\right)^{[z_j]}\left(\frac{n}{n_i}\right)^{-y}\left(\frac{n}{n_i}\right)^{y-z}\right] \\
&= h_{2jzy}k_{6z0}\left(\frac{n}{n_i}\right)^{-z}\left[A_j^{[z_j]}V^{[0]} - \frac{k_{2jzy}k_{4jz0}}{k_{6z0}}\left(A_jV^{[0]}\right)^{[z_j]}\right]
\end{aligned}
\tag{50}
$$

$$
\begin{aligned}
R_{7zy} &= h_{7zy}\left[I^{[z]}V^{[y]} - k_{7zy}\left(\frac{n}{n_i}\right)^{-(y+z)}\right] \\
&= h_{7zy}\left[k_{5z0}I^{[0]}\left(\frac{n}{n_i}\right)^{-z}k_{6y0}V^{[0]}\left(\frac{n}{n_i}\right)^{-y} - k_{7zy}\left(\frac{n}{n_i}\right)^{-(y+z)}\right] \\
&= h_{7zy}k_{5z0}k_{6y0}\left(\frac{n}{n_i}\right)^{-(y+z)}\left[I^{[0]}V^{[0]} - \frac{k_{7zy}}{k_{5z0}k_{6y0}}\right]
\end{aligned}
\tag{51}
$$

$$
\begin{aligned}
R_{8jzy} &= h_{8jzy}\left[\left(A_jI\right)^{[z_j+z]}V^{[y]} - k_{8jzy}A_j^{[z_j]}\left(\frac{n}{n_i}\right)^{-(y+z)}\right] \\
&= h_{8jzy}\left[k_{3jz0}\left(A_jI^{[0]}\right)^{[z_j]}\left(\frac{n}{n_i}\right)^{-z}k_{6y0}V^{[0]}\left(\frac{n}{n_i}\right)^{-y} - k_{8jzy}A_j^{[z_j]}\left(\frac{n}{n_i}\right)^{-(y+z)}\right] \\
&= h_{8jzy}k_{3jz0}k_{6y0}\left(\frac{n}{n_i}\right)^{-(y+z)}\left[\left(A_jI^{[0]}\right)^{[z_j]}V^{[0]} - \frac{k_{8jzy}}{k_{3jz0}k_{6y0}}A_j^{[z_j]}\right]
\end{aligned}
\tag{52}
$$

$$
\begin{aligned}
R_{9jzy} &= h_{9jzy}\left[\left(A_jV\right)^{[z_j+z]}I^{[y]} - k_{9jzy}A_j^{[z_j]}\left(\frac{n}{n_i}\right)^{-(y+z)}\right]I^{(z)} \\
&= h_{9jzy}\left[k_{4jz0}\left(A_jV^{[0]}\right)^{[z_j]}\left(\frac{n}{n_i}\right)^{-z}k_{5y0}I^{[0]}\left(\frac{n}{n_i}\right)^{-y} - k_{9jzy}A_j^{[z_j]}\left(\frac{n}{n_i}\right)^{-(y+z)}\right] \\
&= h_{9jzy}k_{4jz0}k_{5y0}\left(\frac{n}{n_i}\right)^{-(y+z)}\left[\left(A_jV^{[0]}\right)^{[z_j]}I^{[0]} - \frac{k_{9jzy}}{k_{4jz0}k_{5y0}}A_j^{[z_j]}\right]
\end{aligned}
\tag{53}
$$

THERMAL EQUILIBRIUM IN IMPURITY AND POINT DEFECT PARING

We further assume that pairing between impurity and point defects are in thermal equilibrium. Therefore, we assume that $R_{1jzy} = R_{2jzy} = 0$. This assumptions are not well justified, but simplify the 5 stream model significantly, and the resultant model is commonly implemented in commercial simulator.

Under these assumptions, we can relate a pairing concentration to the impurity and point defects concentrations as below.

$$R_{1jzy} = h_{1jzy} k_{5z0} \left(\frac{n}{n_i} \right)^{-z} \left[A_j^{[z_j]} I^{[0]} - \frac{k_{1jzy} k_{3jz0}}{k_{5z0}} \left(A_j I^{[0]} \right)^{[z_j]} \right] = 0 \tag{54}$$

We then obtain

$$\left(A_j I^{[0]} \right)^{[z_j]} = \frac{k_{5z0}}{k_{1jzy} k_{3jz0}} A_j^{[z_j]} I^{[0]} \tag{55}$$

$$R_{2jzy} = h_{2jzy} k_{6z0} \left(\frac{n}{n_i} \right)^{-z} \left[A_j^{[z_j]} V^{[0]} - \frac{k_{2jzy} k_{4jz0}}{k_{6z0}} \left(A_j V^{[0]} \right)^{[z_j]} \right] = 0 \tag{56}$$

We then obtain

$$\left(A_j V^{[0]} \right)^{[z_j]} = \frac{k_{6z0}}{k_{2jzy} k_{4jz0}} A_j^{[z_j]} V^{[0]} \tag{57}$$

In this case, we can focus on individual total species concentrations, that is

$$I_{tot} = \sum_z \left[I^{[z]} + \left(A_j I \right)^{[z_j+z]} \right] \tag{58}$$

$$V_{tot} = \sum_z \left[V^{[z]} + \left(A_j V \right)^{[z_j+z]} \right] \tag{59}$$

$$A_j^{[z_j]}{}_{tot} = A_j^{[z_j]} + \sum_z \left[\left(A_j I \right)^{[z_j+z]} + \left(A_j V \right)^{[z_j+z]} \right] \tag{60}$$

Therefore, we summarize the balance equations as follows.

I :

$$\frac{\partial}{\partial t}\sum_z\left[I^{[z]}+\left(A_jI\right)^{[z_j+z]}\right]+div\sum_z\left[J_{I^{[z]}}+J_{\left(A_jI\right)^{[z_j+z]}}\right]$$

$$=-\sum_{zy}R_{7zy}-\sum_{zy}R_{8jzy}-\sum_{jzy}R_{9jzy}$$

(61)

where

$$\sum_z\left[J_{I^{[z]}}+J_{\left(A_jI\right)^{[z_j+z]}}\right]$$

$$=-\sum_z\left[k_{5z0}D_{I^{[z]}}\left(\frac{n}{n_i}\right)^{-z}\nabla I^{[0]}\right]$$

$$-\sum_z\left[k_{3jz0}D_{\left(A_jI\right)^{[z_j+z]}}\left(\frac{n}{n_i}\right)^{-z}\left\{\nabla\left(A_jI^{[0]}\right)^{[z_j]}+z_j\left(A_jI^{[0]}\right)^{[z_j]}\nabla\left[\ln\left(\frac{n}{n_i}\right)\right]\right\}\right]$$

$$=-\sum_z\left[k_{5z0}D_{I^{[z]}}\left(\frac{n}{n_i}\right)^{-z}\nabla I^{[0]}\right]$$

$$-\sum_z\left[\frac{k_{5z0}}{k_{1jzy}}D_{\left(A_jI\right)^{[z_j+z]}}\left(\frac{n}{n_i}\right)^{-z}\left\{\nabla\left[A_j^{[z_j]}I^{[0]}\right]+z_j\left[A_j^{[z_j]}I^{[0]}\right]\nabla\left[\ln\left(\frac{n}{n_i}\right)\right]\right\}\right]$$

(62)

V :

$$\frac{\partial}{\partial t}\sum_z\left[V^{[z]}+\left(A_jV\right)^{[z_j+z]}\right]+div\sum_z\left[J_{V^{[z]}}+J_{\left(A_jV\right)^{[z_j+z]}}\right]$$

$$=-\sum_{zy}R_{7yz}-\sum_{jzy}R_{8jyz}-\sum_{zy}R_{9jzy}$$

(63)

where

$$\sum_z \left[J_{V^{[z]}} + J_{\left(A_j V\right)^{[z_j+z]}} \right]$$

$$= -\sum_z k_{6z0} D_{V^{[z]}} \left(\frac{n}{n_i}\right)^{-z} \nabla V^{[0]}$$

$$- \sum_z k_{4jz0} D_{\left(A_j V\right)^{[z_j+z]}} \left(\frac{n}{n_i}\right)^{-z} \left\{ \nabla \left(A_j V^{[0]}\right)^{[z_j]} + z_j \left(A_j V^{[0]}\right)^{[z_j]} \nabla \left[\ln\left(\frac{n}{n_i}\right)\right] \right\}$$

$$= -\sum_z k_{6z0} D_{V^{[z]}} \left(\frac{n}{n_i}\right)^{-z} \nabla V^{[0]}$$

$$- \sum_z \frac{k_{6z0}}{k_{2jzy}} D_{\left(A_j V\right)^{[z_j+z]}} \left(\frac{n}{n_i}\right)^{-z} \left\{ \nabla \left[A_j^{[z_j]} V^{[0]} \right] + z_j \left[A_j^{[z_j]} V^{[0]} \right] \nabla \left[\ln\left(\frac{n}{n_i}\right)\right] \right\} \tag{64}$$

$A_j:$

$$\frac{\partial}{\partial t} \sum_z \left[A_j^{[z_j]} + \left(A_j I\right)^{[z_j+z]} + \left(A_j V\right)^{[z_j+z]} \right] + div \sum_z \left[J_{\left(A_j I\right)^{[z_j+z]}} + J_{\left(A_j V\right)^{[z_j+z]}} \right] = 0 \tag{65}$$

where

$$\sum_z \left[J_{\left(A_j I\right)^{[z_j+z]}} + J_{\left(A_j V\right)^{[z_j+z]}} \right]$$

$$= -\sum_z k_{3jz0} D_{\left(A_j I\right)^{[z_j+z]}} \left(\frac{n}{n_i}\right)^{-z} \left\{ \nabla \left(A_j I^{[0]}\right)^{[z_j]} + z_j \left(A_j I^{[0]}\right)^{[z_j]} \nabla \left[\ln\left(\frac{n}{n_i}\right)\right] \right\}$$

$$- \sum_z k_{4jz0} D_{\left(A_j V\right)^{[z_j+z]}} \left(\frac{n}{n_i}\right)^{-z} \left\{ \nabla \left(A_j V^{[0]}\right)^{[z_j]} + z_j \left(A_j V^{[0]}\right)^{[z_j]} \nabla \left[\ln\left(\frac{n}{n_i}\right)\right] \right\}$$

$$= -\sum_z \frac{k_{5z0}}{k_{1jzy}} D_{\left(A_j I\right)^{[z_j+z]}} \left(\frac{n}{n_i}\right)^{-z} \left\{ \nabla \left[A_j^{[z_j]} I^{[0]} \right] + z_j \left[A_j^{[z_j]} I^{[0]} \right] \nabla \left[\ln\left(\frac{n}{n_i}\right)\right] \right\}$$

$$- \sum_z \frac{k_{6z0}}{k_{2jzy}} D_{\left(A_j V\right)^{[z_j+z]}} \left(\frac{n}{n_i}\right)^{-z} \left\{ \nabla \left[A_j^{[z_j]} V^{[0]} \right] + z_j \left[A_j^{[z_j]} V^{[0]} \right] \nabla \left[\ln\left(\frac{n}{n_i}\right)\right] \right\} \tag{66}$$

We should solve coupled Eqs. of 61, 63, and 65, which is called as 3 stream model.

Under these approximations, each recombination rate is simplified as below.

$$R_{7zy} = h_{7zy} k_{5z0} k_{6y0} \left(\frac{n}{n_i} \right)^{-(y+z)} \left[I^{[0]} V^{[0]} - \frac{k_{7zy}}{k_{5z0} k_{6y0}} \right] \tag{67}$$

$$R_{8jzy} = h_{8jzy} k_{3jz0} k_{6y0} \left(\frac{n}{n_i} \right)^{-(y+z)} \left[\left(A_j I^{[0]} \right)^{[z_j]} V^{[0]} - \frac{k_{8jzy}}{k_{3jz0} k_{6y0}} A_j^{[z_j]} \right]$$

$$= h_{8jzy} k_{3jz0} k_{6y0} \left(\frac{n}{n_i} \right)^{-(y+z)} \left[\frac{k_{5z0}}{k_{1jzy} k_{3jz0}} A_j^{[z_j]} I^{[0]} V^{[0]} - \frac{k_{8jzy}}{k_{3jz0} k_{6y0}} A_j^{[z_j]} \right]$$

$$= h_{8jzy} k_{6y0} \frac{k_{5z0}}{k_{1jzy}} \left(\frac{n}{n_i} \right)^{-(y+z)} A_j^{[z_j]} \left[I^{[0]} V^{[0]} - \frac{k_{8jzy} k_{1jzy}}{k_{6y0} k_{5z0}} \right] \tag{68}$$

$$R_{9jzy} = h_{9jzy} k_{4jz0} k_{5y0} \left(\frac{n}{n_i} \right)^{-(y+z)} \left[\left(A_j V^{[0]} \right)^{[z_j]} I^{[0]} - \frac{k_{9jzy}}{k_{4jz0} k_{5y0}} A_j^{[z_j]} \right]$$

$$= h_{9jzy} k_{4jz0} k_{5y0} \left(\frac{n}{n_i} \right)^{-(y+z)} \left[\frac{k_{6z0}}{k_{2jzy} k_{4jz0}} A_j^{[z_j]} V^{[0]} I^{[0]} - \frac{k_{9jzy}}{k_{4jz0} k_{5y0}} A_j^{[z_j]} \right]$$

$$= h_{9jzy} k_{5y0} \frac{k_{6z0}}{k_{2jzy}} \left(\frac{n}{n_i} \right)^{-(y+z)} A_j^{[z_j]} \left[V^{[0]} I^{[0]} - \frac{k_{9jzy} k_{2jzy}}{k_{5y0} k_{6z0}} \right] \tag{69}$$

THERMAL EQUILIBRIUM IN POINT DEFECTS RECOMBINATION

We further approximate that the recombination of point defects are also in thermal equilibrium. Under this condition, R_{7zy}, R_{8jzy} and R_{9jzy} can be set 0. Therefore, we can approximately express the equations as

$$R_{7zy} = h_{7zy} k_{5z0} k_{6y0} \left(\frac{n}{n_i} \right)^{-(y+z)} \left[I^{[0]} V^{[0]} - I^{*[0]} V^{*[0]} \right] \tag{70}$$

$$R_{8jzy} = h_{8jzy} k_{6y0} \frac{k_{5z0}}{k_{1jzy}} \left(\frac{n}{n_i} \right)^{-(y+z)} A_j^{[z_j]} \left[I^{[0]} V^{[0]} - I^{*[0]} V^{*[0]} \right] \tag{71}$$

$$R_{9jzy} = h_{9jzy} k_{5y0} \frac{k_{6z0}}{k_{2jzy}} \left(\frac{n}{n_i} \right)^{-(y+z)} A_j^{[z_j]} \left[I^{[0]} V^{[0]} - I^{*[0]} V^{*[0]} \right] \tag{72}$$

where $I^{*[0]}$ and $V^{*[0]}$ are neutral interstitial Si and vacancy concentrations in thermal equilibrium, respectively.

In the early stage of diffusion, the species recombine significantly, and we should treat them dynamically. However, after a certain time period, the recombination of point defects is not so significant in the bulk, and point defects mainly recombine at the surface. If we set initial condition of diffusion process as this stage, this approximation may become valid. Under this approximation, we can impose

$$I^{[0]} V^{[0]} = I^{*[0]} V^{*[0]} \tag{73}$$

The three stream models are then reduced to as follows.

$$\frac{\partial}{\partial t} \sum_z \left[I^{[z]} + \left(A_j I \right)^{[z_j + z]} \right] + div \sum_z \left[J_{I^{[z]}} + J_{\left(A_j I \right)^{[z_j + z]}} \right] = 0 \tag{74}$$

where

$$\sum_z \left[J_{I^{[z]}} + J_{\left(A_j I \right)^{[z_j + z]}} \right]$$

$$= -\sum_z \left[k_{5z0} D_{I^{[z]}} \left(\frac{n}{n_i} \right)^{-z} \nabla I^{[0]} \right]$$

$$- \sum_z \left[\frac{k_{5z0}}{k_{1jzy}} D_{\left(A_j I \right)^{[z_j + z]}} \left(\frac{n}{n_i} \right)^{-z} \left\{ \nabla \left[A_j^{[z_j]} I^{[0]} \right] + z_j \left[A_j^{[z_j]} I^{[0]} \right] \nabla \left[\ln \left(\frac{n}{n_i} \right) \right] \right\} \right] \tag{75}$$

$$\frac{\partial}{\partial t} \sum_z \left[V^{[z]} + \left(A_j V \right)^{[z_j + z]} \right] + div \sum_z \left[J_{V^{[z]}} + J_{\left(A_j V \right)^{[z_j + z]}} \right] = 0 \tag{76}$$

where

$$\sum_z \left[J_{V^{[z]}} + J_{(A_jV)^{[z_j+z]}} \right]$$

$$= -\sum_z k_{6z0} D_{V^{[z]}} \left(\frac{n}{n_i} \right)^{-z} \nabla V^{[0]}$$

$$-\sum_z \frac{k_{6z0}}{k_{2jzy}} D_{(A_jV)^{[z_j+z]}} \left(\frac{n}{n_i} \right)^{-z} \left\{ \nabla \left[A_j^{[z_j]} V^{[0]} \right] + z_j \left[A_j^{[z_j]} V^{[0]} \right] \nabla \left[\ln \left(\frac{n}{n_i} \right) \right] \right\} \tag{77}$$

$$\frac{\partial}{\partial t} \sum_z \left[A_j^{[z_j]} + \left(A_jI \right)^{[z_j+z]} + \left(A_jV \right)^{[z_j+z]} \right] + div \sum_z \left[J_{(A_jI)^{[z_j+z]}} + J_{(A_jV)^{[z_j+z]}} \right] = 0 \tag{78}$$

where

$$\sum_z \left[J_{(A_jI)^{[z_j+z]}} + J_{(A_jV)^{[z_j+z]}} \right]$$

$$= -\sum_z \frac{k_{5z0}}{k_{1jzy}} D_{(A_jI)^{[z_j+z]}} \left(\frac{n}{n_i} \right)^{-z} \left\{ \nabla \left[A_j^{[z_j]} I^{[0]} \right] + z_j \left[A_j^{[z_j]} I^{[0]} \right] \nabla \left[\ln \left(\frac{n}{n_i} \right) \right] \right\} \tag{79}$$

$$-\sum_z \frac{k_{6z0}}{k_{2jzy}} D_{(A_jV)^{[z_j+z]}} \left(\frac{n}{n_i} \right)^{-z} \left\{ \nabla \left[A_j^{[z_j]} V^{[0]} \right] + z_j \left[A_j^{[z_j]} V^{[0]} \right] \nabla \left[\ln \left(\frac{n}{n_i} \right) \right] \right\}$$

FURTHER APPROXIMATION

Impurity point defect pair concentration is much lower than those of lattice site impurity concentration in general, and hence we make further approximation.

$$\frac{\partial}{\partial t} \sum_z \left[I^{[z]} + \left(A_jI \right)^{[z_j+z]} \right] + div \sum_z \left[J_{I^{[z]}} + J_{(A_jI)^{[z_j+z]}} \right]$$

$$\approx \frac{\partial}{\partial t} \sum_z I^{[z]} + div \sum_z J_{I^{[z]}} \tag{80}$$

$$= \frac{\partial}{\partial t} \sum_z I^{[z]} - div \sum_z k_{5z0} D_{I^{[z]}} \left(\frac{n}{n_i} \right)^{-z} \nabla I^{[0]}$$

We then obtain

$$\frac{\partial}{\partial t} I_{tot} = div \sum_z k_{5z0} D_{I^{[z]}} \left(\frac{n}{n_i}\right)^{-z} \nabla I^{[0]} \tag{81}$$

Diffusion coefficient of interstitial Si depends on charge state in general. However, we do not know detail of charged dependent diffusion coefficients, and hence we use average one and express it as

$$\frac{\partial}{\partial t} I_{tot} \simeq D_I \nabla^2 I_{tot} \tag{82}$$

Similarly, we obtain

$$\frac{\partial}{\partial t} V_{tot} \approx D_V \nabla^2 V_{tot} \tag{83}$$

We obtain equations associated with impurity as

$$\frac{\partial}{\partial t} A_{jtot}^{[z_j]} + div \sum_z \left[J_{(A_j I)^{[z_j+z]}} + J_{(A_j V)^{[z_j+z]}} \right] = 0 \tag{84}$$

$$\sum_z \left[J_{(A_j I)^{[z_j+z]}} + J_{(A_j V)^{[z_j+z]}} \right]$$

$$= -\sum_z \frac{k_{5z0}}{k_{1jzy}} D_{(A_j I)^{[z_j+z]}} \left(\frac{n}{n_i}\right)^{-z} \left\{ \nabla \left[A_j^{[z_j]} I^{[0]} \right] + z_j \left[A_j^{[z_j]} I^{[0]} \right] \nabla \left[\ln\left(\frac{n}{n_i}\right) \right] \right\}$$

$$-\sum_z \frac{k_{6z0}}{k_{2jzy}} D_{(A_j V)^{[z_j+z]}} \left(\frac{n}{n_i}\right)^{-z} \left\{ \nabla \left[A_j^{[z_j]} V^{[0]} \right] + z_j \left[A_j^{[z_j]} V^{[0]} \right] \nabla \left[\ln\left(\frac{n}{n_i}\right) \right] \right\}$$

$$\approx -\sum_z \frac{k_{5z0}}{k_{1jzy}} D_{(A_j I)^{[z_j+z]}} I^{[0]} \left(\frac{n}{n_i}\right)^{-z} \left\{ \nabla A_j^{[z_j]} + z_j A_j^{[z_j]} \nabla \left[\ln\left(\frac{n}{n_i}\right) \right] \right\}$$

$$-\sum_z \frac{k_{6z0}}{k_{2jzy}} D_{(A_j V)^{[z_j+z]}} V^{[0]} \left(\frac{n}{n_i}\right)^{z} \left\{ \nabla A_j^{[z_j]} + z_j A_j^{[z_j]} \nabla \left[\ln\left(\frac{n}{n_i}\right) \right] \right\}$$

$$= -\left\{ \sum_z \left[\frac{k_{5z0}}{k_{1jzy}} D_{(A_j I)^{[z_j+z]}} I^{[0]} \frac{I^{[0]}}{I^{*[0]}} \left(\frac{n}{n_i}\right)^{-z} + \frac{k_{6z0}}{k_{2jzy}} D_{(A_j V)^{[z_j+z]}} V^{[0]} \frac{V^{[0]}}{V^{*[0]}} \left(\frac{n}{n_i}\right)^{-z} \right] \right\}$$

$$\times \left\{ \nabla A_{jtot}^{[z_j]} + z_j A_{jtot}^{[z_j]} \nabla \left[\ln\left(\frac{n}{n_i}\right) \right] \right\} \tag{85}$$

In the derivation process of Eq. 85, we assume the gradient of point defects are quite low, since their diffusion coefficient is much higher than the pair diffusion coefficient, and hence approximate as

$$\nabla\left[A_j^{[z_j]}I^{[0]}\right] \approx I^{[0]}\nabla A_j^{[z_j]}, \nabla\left[A_j^{[z_j]}V^{[0]}\right] \approx V^{[0]}\nabla A_j^{[z_j]}$$

The diffusion coefficient for each charge state is then given by

$$\frac{k_{5z0}}{k_{1jzy}}D_{(A_jI)^{[z_j+z]}}I^{*[0]}\frac{I^{[0]}}{I^{*[0]}}\left(\frac{n}{n_i}\right)^{-z} + \frac{k_{6z0}}{k_{2jzy}}D_{(A_jV)^{[z_j+z]}}V^{*[0]}\frac{V^{[0]}}{V^{*[0]}}\left(\frac{n}{n_i}\right)^{-z}$$

$$= \left[\frac{k_{5z0}}{k_{1jzy}}D_{(A_jI)^{[z_j+z]}}I^{*[0]} + \frac{k_{6z0}}{k_{2jzy}}D_{(A_jV)^{[z_j+z]}}V^{*[0]}\right]\left(\frac{n}{n_i}\right)^{-z}$$

$$\times \left[\frac{\dfrac{k_{5z0}}{k_{1jzy}}D_{(A_jI)^{[z_j+z]}}I^{*[0]}}{\dfrac{k_{5z0}}{k_{1jzy}}D_{(A_jI)^{[z_j+z]}}I^{*[0]} + \dfrac{k_{6z0}}{k_{2jzy}}D_{(A_jV)^{[z_j+z]}}V^{*[0]}}\frac{I^{[0]}}{I^{*[0]}} + \frac{\dfrac{k_{6z0}}{k_{2jzy}}D_{(A_jV)^{[z_j+z]}}V^{*[0]}}{\dfrac{k_{5z0}}{k_{1jzy}}D_{(A_jI)^{[z_j+z]}}I^{*[0]} + \dfrac{k_{6z0}}{k_{2jzy}}D_{(A_jV)^{[z_j+z]}}V^{*[0]}}\frac{V^{[0]}}{V^{*[0]}}\right]$$

$$= D^{[z]}\left(\frac{n}{n_i}\right)^{-z}\left[f_I\frac{I^{[0]}}{I^{*[0]}} + (1-f_I)\frac{V^{[0]}}{V^{*[0]}}\right] \tag{86}$$

where

$$D^{[z]} = \frac{k_{5z0}}{k_{1jzy}}D_{(A_jI)^{[z_j+z]}}I^{*[0]} + \frac{k_{6z0}}{k_{2jzy}}D_{(A_jV)^{[z_j+z]}}V^{*[0]} \tag{87}$$

$$f_I = \frac{\dfrac{k_{5z0}}{k_{1jzy}}D_{(A_jI)^{[z_j+z]}}I^{*[0]}}{\dfrac{k_{5z0}}{k_{1jzy}}D_{(A_jI)^{[z_j+z]}}I^{*[0]} + \dfrac{k_{6z0}}{k_{2jzy}}D_{(A_jV)^{[z_j+z]}}V^{*[0]}} \tag{88}$$

We further replace the neutral point defect concentrations to the total ones as

$$\frac{\partial}{\partial t}A_{jtot}^{[z_j]} = div\sum_z D^{[z]}\left(\frac{n}{n_i}\right)^{-z}\left[f_I\frac{I}{I^*} + (1-f_I)\frac{V}{V^*}\right]\nabla A_{jtot}^{[z_j]} \tag{89}$$

where we express total ones without no charge states. This is the equation derived in the previous section.

We assume that I and V are not independent, but are related to

$$IV = I^*V^* \tag{90}$$

as was assumed for neutral ones. Focusing on interstitial Si, we modify Eq. 89 as

$$\frac{\partial}{\partial t} A_{jtot}{}^{[z_j]} = div \sum_z D^{[z]} \left(\frac{n}{n_i}\right)^{-z} \left[f_I \frac{I}{I^*} + \left(1 - f_I\right) \frac{I^*}{I} \right] \nabla A_{jtot}{}^{[z_j]} \tag{91}$$

Finally, we can describe diffusion profiles using Eqs. 82 and 91.

REFERENCES

[1]   P. Pichler, Intrinsic Point Defects, Impurities, and Their Diffusion in Silicon, Springer-Verlag/Wien, 2004, Chapter 5.
[2]   DIOS users' manual.

Send Orders for Reprints to reprints@benthamscience.net

CHAPTER 3

Experimental Data Associated with Transient Diffusion

Abstract: We should determine various ambiguous physical parameters of the transient diffusion (TED) models associated with paring to understand TED fully. On the other hand, we can only obtain total impurity concentration profile data which lack paring information. We should guess the paring parameters from total redistributed impurity concentration profiles. The difficulty that fundamental data can only be evaluated at low-temperature regions also exist, that is, we cannot simply relate the profiles to a target temperature set at practical temperatures of about 1000°C. This means that the TED starts and ends in the ramping process at the target temperature. Therefore, we show TED data at low temperatures, and clarify its prominent features. The characteristics at high temperatures can be evaluated by extrapolating the parameter values.

Keywords: Diffusion, transient enhanced diffusion, oxidation, point defects, ion implantation, interstitial Si, vacancy, rapid annealing, ramp up rate, redistribution, shallow junction, VLSI process, amorphous/crystal interface, diffusion length, clusters, solid solubility, B marker layer.

INTRODUCTION

We observe transient diffusion at early stages of diffusion or in oxidation process. The reason for this is that there are the excess point defects induced by ion implantation or oxidation. Impurities which are apt to form pairs with interstitial Si show transient enhanced diffusion (TED) and the ones with vacancy show transient retarded diffusion. The former corresponds to B and In, and the latter is corresponding to Sb. Significant work has been done on this subject [1-7].

Although the microscopic understanding of TED is difficult, the macroscopic prominent features of TED are rather simple, which are

(a) The diffusion coefficient is quite higher than that under the thermal equilibrium by many orders.

(b) TED occurs and finishes in quite a short time.

(c) There are maximum concentrations where TED occurs.

(a) is related to point defects that are introduced by many orders and higher than those in the thermal equilibrium with ion implantation or oxidation.

(b) is related to diffusion coefficients of point defects, which are quite higher than those of impurities.

(c) is not clear which it is related to.

We summarize systematic experimental data associated with TED reported in [8-13], add some in this chapter, and clarify prominent features of TED.

DEPENDENCE OF B TRANSIENT ENHANCED DIFFUSION ON TEMPERATURE

Fig. **1** shows B diffusion profiles after rapid thermal annealing (RTA) for 10 sec. The ramp up rate was 50°C/sec. We do not observe redistribution of the as implanted profiles up to the temperatures of 650°C, and observe redistributions at higher temperatures.

The diffusion profiles in the tail low concentration region are almost the same at temperatures higher than 850°C. It takes a few seconds during the ramp up process between 850°C to 1000°C. The same profiles in the low concentration region indicate that TED starts and finishes during this ramp up time period. This means that TED time is of the order of second around 900°C, and that the profiles associated with TED are not related to the target temperatures but to the ramp up process at temperatures higher than 850°C.

Fig. **1** also shows that TED goes on for 10 second at temperatures of less than 800°C. The time period we can control is a few second in RTA and a few minutes in furnace annealing (FA). We can obtain experimental data neglecting the ramp up time period at the temperatures of less than 800°C. Therefore, we mainly focus on the low temperature region, although the typical annealing temperature is around 1000°C in practical VLSI processes. We use RTA when TED time is less than few seconds, and furnace anneal (FA) when the time is much longer than a few minutes to relate the profiles to the target temperatures.

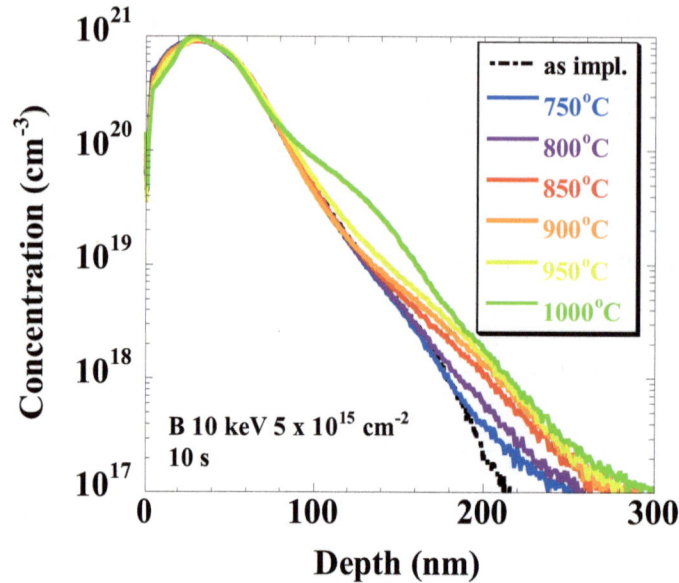

Figure 1: Summary of diffusion profiles at temperatures exceeding 750°C. We observe little diffusion less than these temperatures.

ANNEALING TEMPERATURE AND TIME DEPENDENCE OF B AND In DIFFUSION PROFILES

Figs. **2** and **3** show time dependence of B and In diffusion profiles at various temperatures for various time periods, respectively. Significant diffusion occurs in the low concentration region and only in the early stage of the diffusion and then the profiles become invariable after a certain time period. The B diffusion profiles at 1000°C do not depend on annealing time while those at 800°C do depend on the time in the same time duration. This clearly means that the TED profiles are determined during ramp up time period at 1000°C, and the profiles are hence not related to the target temperature at 1000°C.

We observe kinks in the peak region for In diffusion profiles. This may be related to the trap of In to residual defects which exists at the amorphous/crystal interface. We do not discuss this phenomenon here.

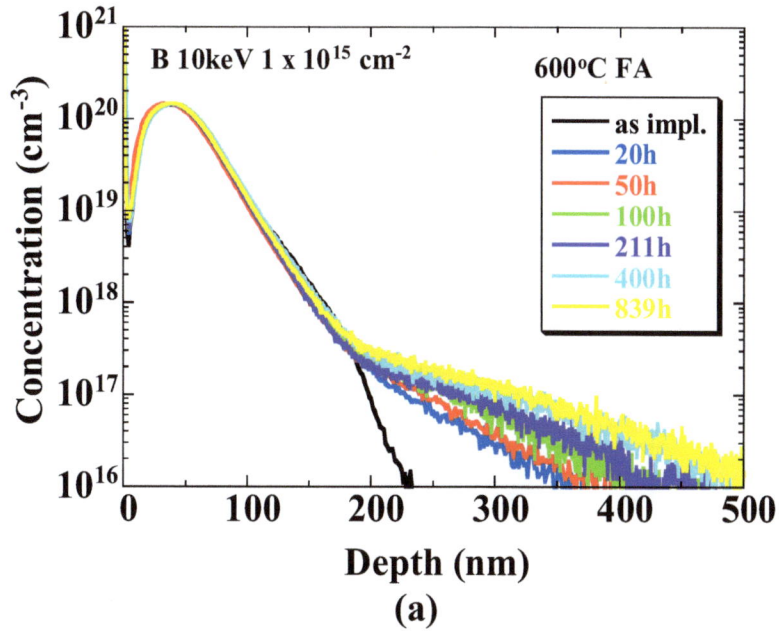

(a)

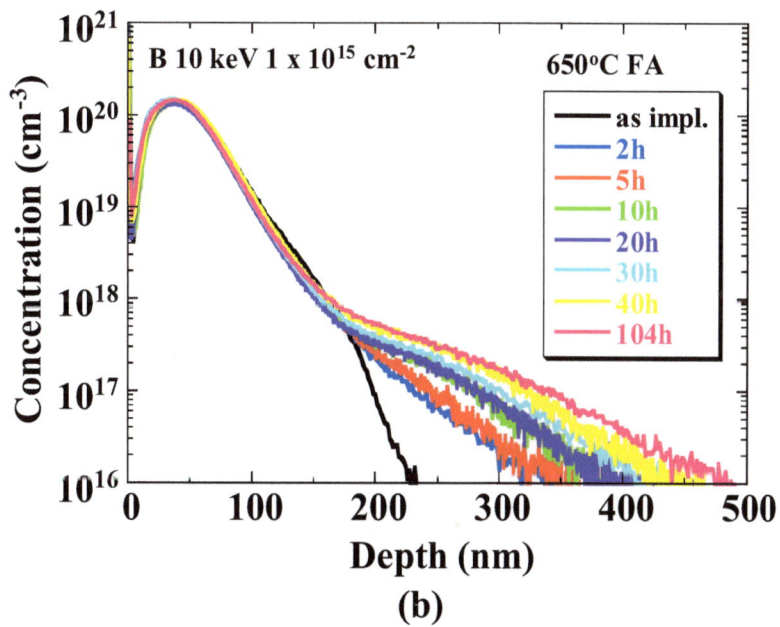

(b)

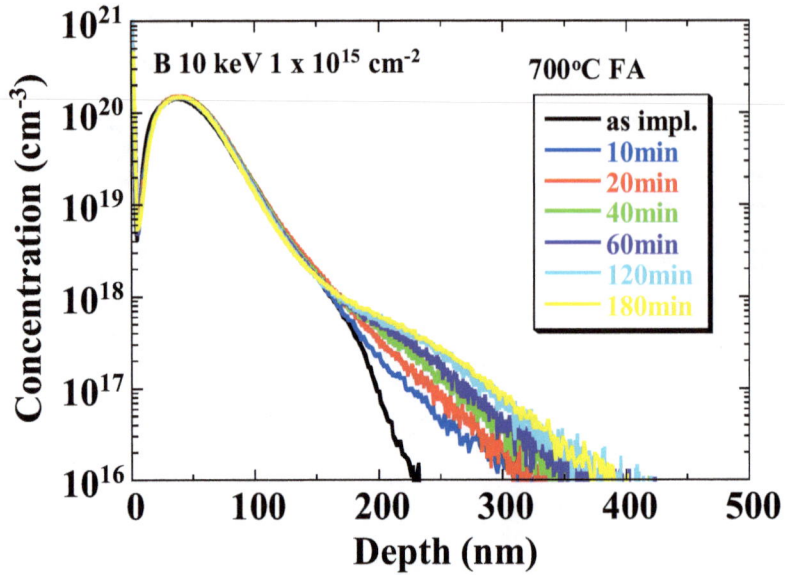

(c)

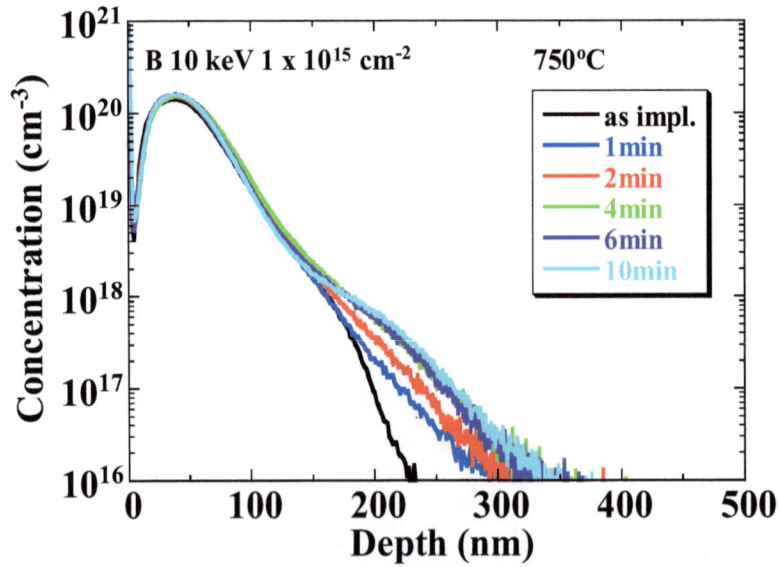

(d)

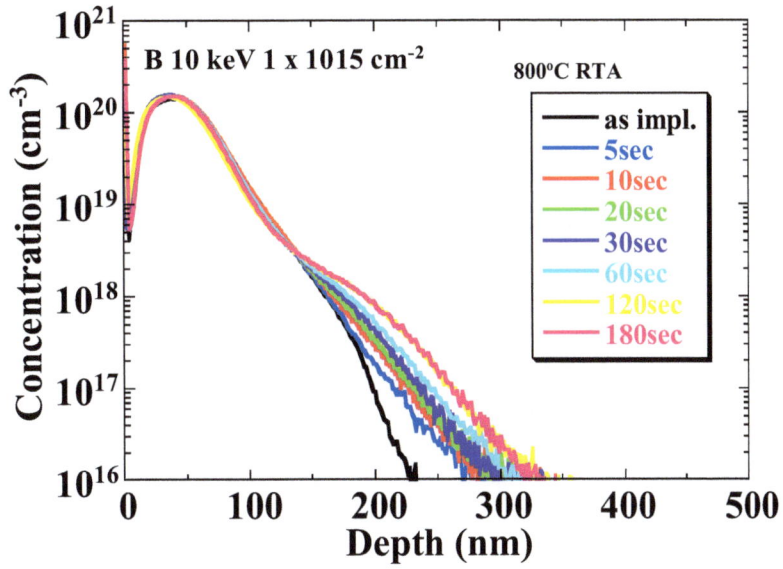

(e)

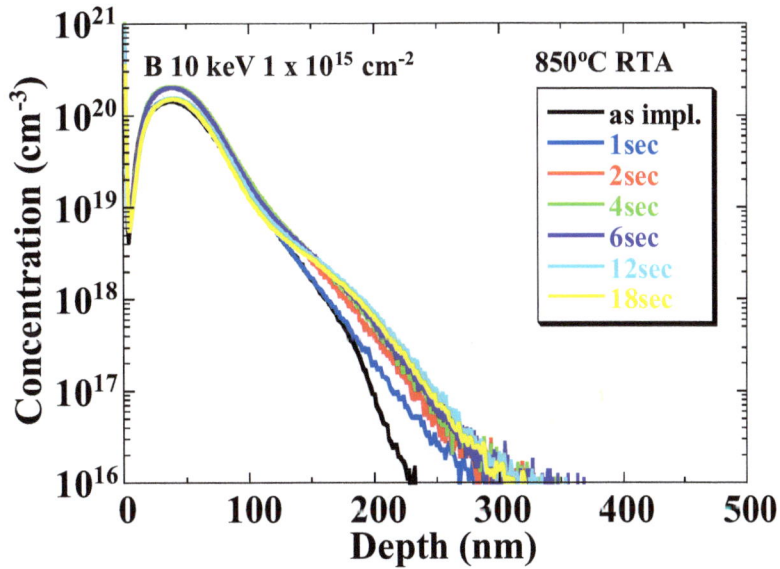

(f)

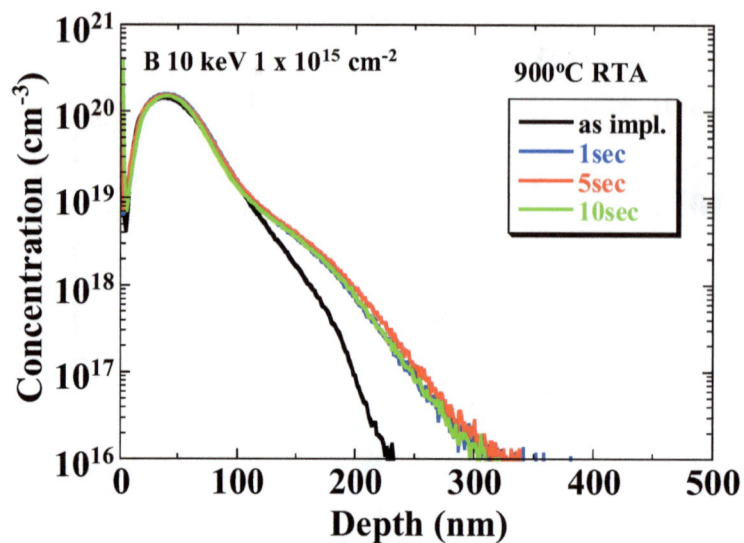

(g)

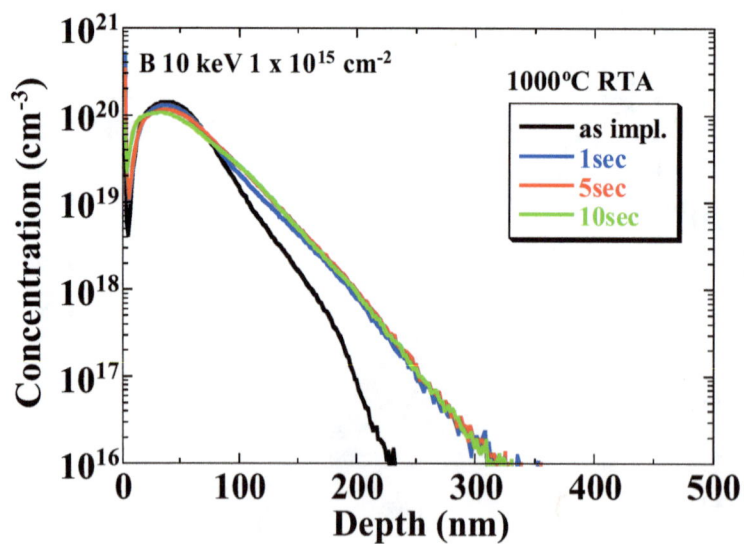

(h)

Figure 2: B diffusion profiles for various temperatures and times. (a) 600°C, (b) 650°C, (c) 700°C, (d) 750°C, (e) 800°C, (f) 850°C, (g) 900°C, (h) 1000°C.

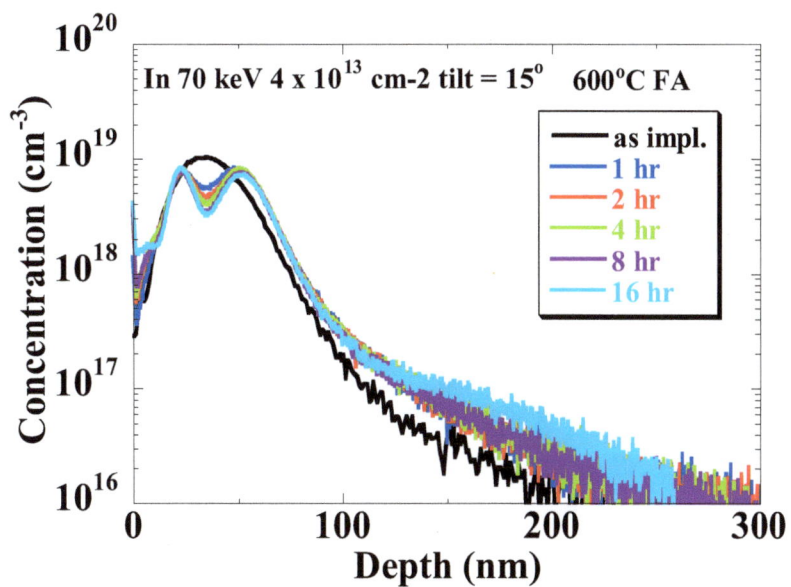

(a)

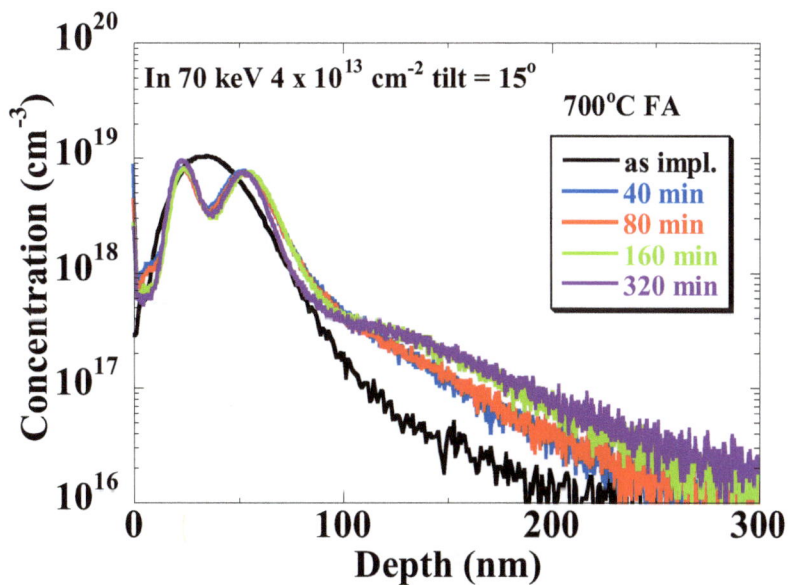

(b)

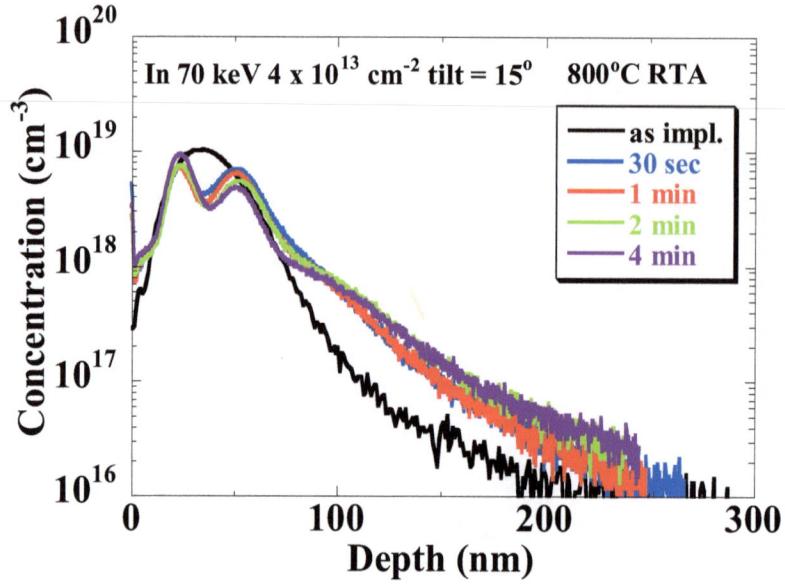

(c)

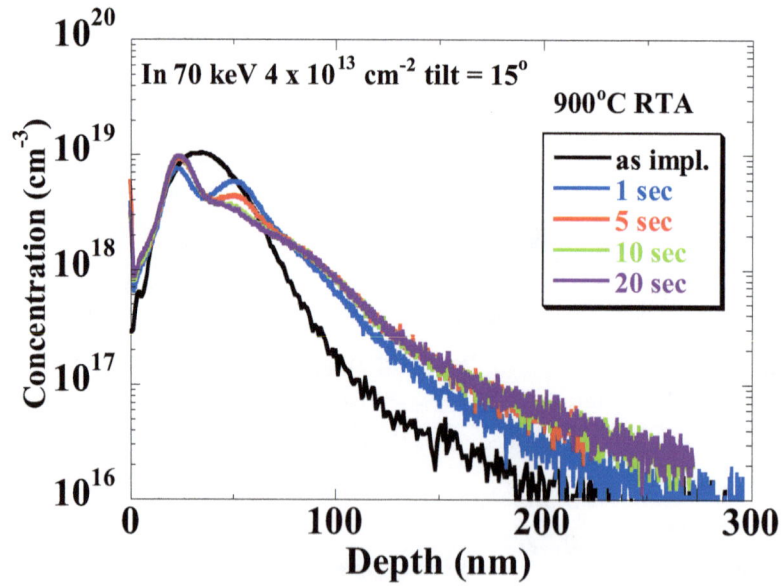

(d)

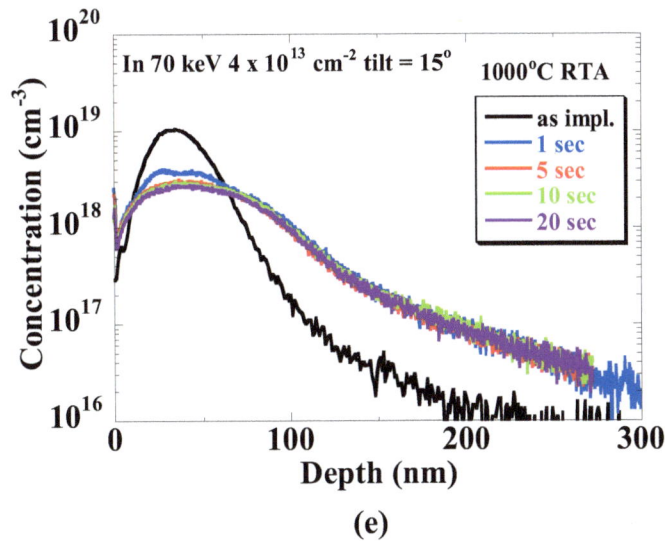

(e)

Figure 3: In diffusion profiles for various time and temperatures. (a) 600°C, (b) 700°C, (c) 800°C, (d) 900°C, (e) 1000°C.

We tried to find appropriate diffusion time where we observe the profiles before and after TED finishing in Figs. **2** and **3**. We can obtain such data at less than 800°C, but it is difficult to obtain such data at around 1000°C where the profiles are determined during ramp up time period as we mentioned. We should set much high ramp up rate to obtain the data in such high temperature region.

Fig. **4** shows the dependence of B diffusion length on time at 600 and 700°C. The diffusion length is evaluated as the distance between the depth where the concentration is 10^{17} cm^{-3} and that of as implanted profile as shown in the inset figure in Fig. **4**(b). The diffusion length is proportional to square root of time, and then saturates. This means that we can regard that the enhanced diffusion coefficient is constant during TED. The gradient at 700°C is higher than that at 600°C, but the saturated diffusion length at 700°C is smaller than that at 600°C.

The interstitial Si is introduced by ion implantation, and it seems that the interstitial Si concentration decreases with time increase during thermal process. Based on the feature, we should expect the decrease of transient enhanced diffusion coefficient. However, we observe constant diffusion coefficient as shown in Fig. **4**. This can be explained as follows.

Interstitial Si induced by ion implantation first form clusters of 311 defects [14-18], and the clusters release the mobile intestinal Si. The clusters play a role of constant concentration interstitial Si source. This is rather complex, but it makes the macroscopic transient enhanced diffusion features very simple, and we observe constant diffusion coefficient during TED.

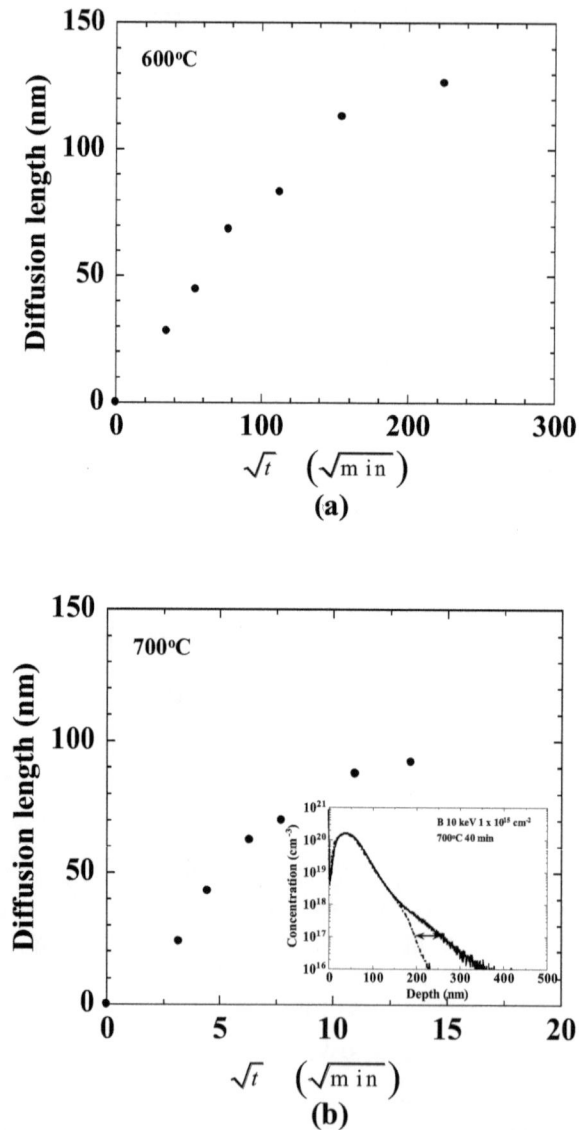

Figure 4: Dependence of diffusion length on time. (a) 600°C, (b) 700°C.

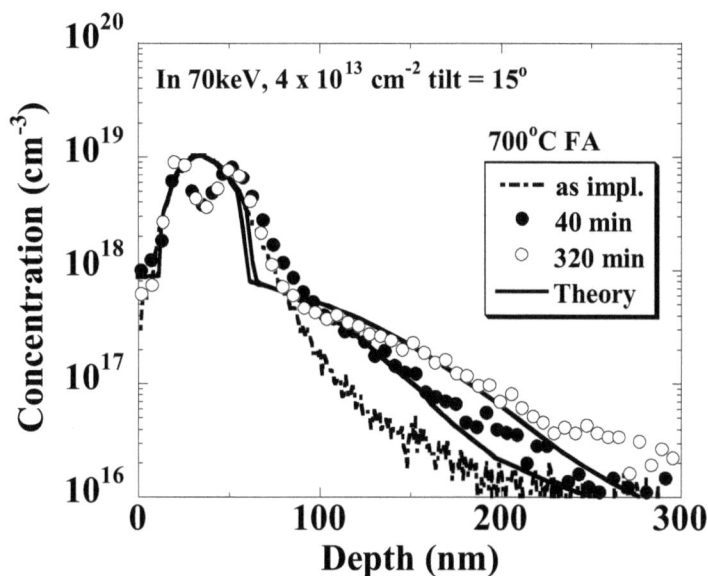

Figure 5: Fitting of simulation to experimental diffusion profiles. The profile for 40 min is the one on the way of TED. Simulation is fitted to the data by changing diffusion coefficient. The profile for 320 min is the one after the completion of TED, and it is invaluable with time. Simulation is fitted to the data by changing time with extracted diffusion coefficient.

We can evaluate D_{enh}, t_{enh}, and maximum diffusion concentration from Figs. **2** and **3** using a commercial simulator with thermal equilibrium diffusion model as the following.

Fig. **5** shows one of the fitted data where we use In data which shows the diffusion at 700°C for 40 and 320 minutes. First, we use data for 40 min where TED is on the way. There is the maximum diffusion concentration during TED as shown in Figs. **2** and **3**. This mechanism is not clear at present. However, we can express it with solid solubility in the simulator. The enhanced diffusion profiles are simply reproduced with fitted diffusion coefficient, which is multiplied by many orders. We can regard the extracted values as enhanced diffusion coefficient D_{enh}.

Next, we use data for 320 min where TED is completed. We fit the theory to the data by changing time with the extracted enhanced diffusion coefficient and the maximum TED diffusion concentration. The fitted time corresponds to t_{enh}.

The fitted profiles for both before and after the completion of TED are shown in Fig. **5**. We performed similar analysis for other temperatures, and obtained temperature dependence of t_{enh} and the maximum diffusion concentration.

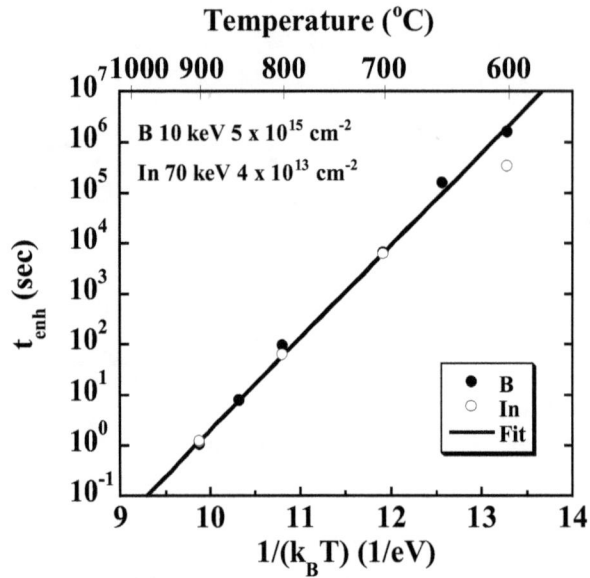

Figure 6: Dependence of extracted transient enhanced diffusion time on temperature.

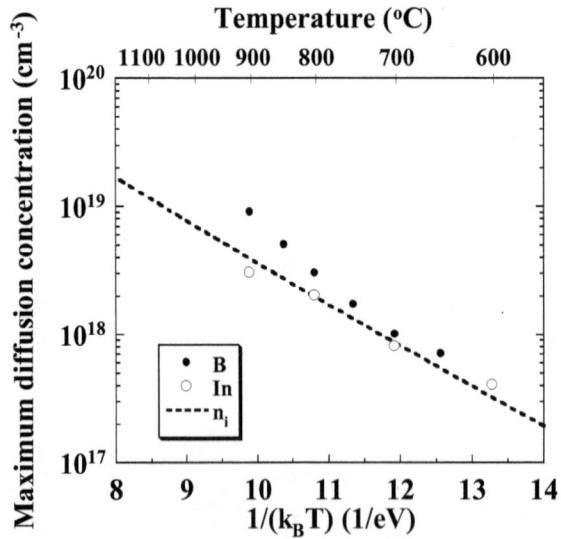

Figure 7: Dependence of extracted maximum diffusion concentration during TED on temperature.

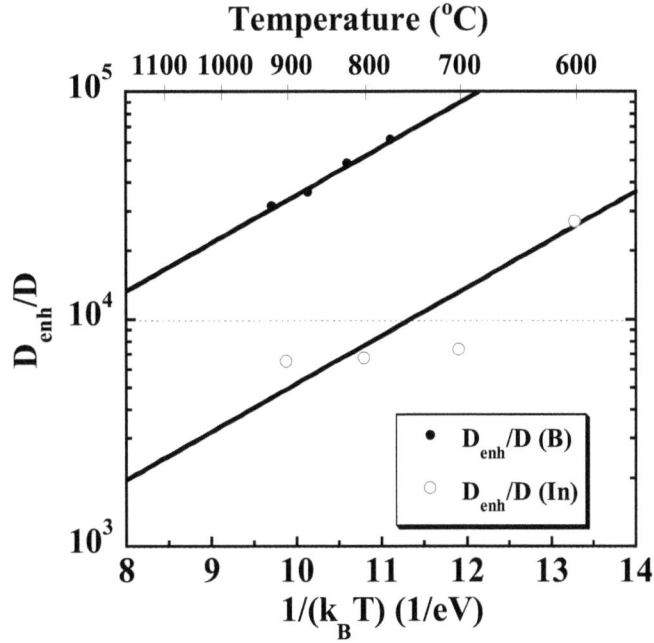

Figure 8: Dependence of enhanced diffusion coefficients on temperature.

Fig. **6** shows the extracted t_{enh} for B and In. It is noteworthy that t_{enh} for both B and In are almost the same. The data are well fitted to the theory and the fitted data has a form with one activation energy given by

$$t_{enh} = 1.06 \times 10^{-18} \exp\left(\frac{4.21(eV)}{k_B T} \right) \sec \tag{1}$$

t_{enh} is about 1 sec at 900°C, and extrapolated t_{enh} is 0.05 sec at 1000°C. Therefore, we cannot observe TED profiles related to the isothermal process at 1000°C as we mentioned before.

Fig. **7** shows the extracted maximum diffusion concentration during TED. It is noted that it is close to the intrinsic carrier concentration. The origin is not clear at present. This was tried to explain Fermi level dependent impurity activation mechanism [8, 10]. However, impurities are activated very quickly as shown in chapter 9 of volume 2, and the other mechanism should be investigated.

Fig. **8** shows the extracted diffusion coefficients D_{enh}, which were extracted with the following procedure.

We assume that the enhanced diffusion coefficient is given by

$$D_{enh} = \left| \frac{[I]}{[I]^*} f_{Ieff} + \frac{[V]}{[V]^*} \left(1 - f_{Ieff}\right) \right| D_i$$

$$\approx \frac{[I]}{[I]^*} f_{Ieff} D_i \tag{2}$$

We assume that $[I]$ are independent of impurity, and that f_{Ieff} depends on impurity, but does not depend on temperature. Therefore, we extract the diffusion coefficient so that the activation energy is the same for both B and In, and obtain

$$\frac{D_{enh}}{D_i}(B) = 270 \exp\left| \frac{0.487(eV)}{k_B T} \right| \tag{3}$$

$$\frac{D_{enh}}{D_i}(In) = 40 \exp\left| \frac{0.487(eV)}{k_B T} \right| \tag{4}$$

where the diffusion coefficients of In [19] and B [20] in thermal equilibrium are given by

$$D_i(In) = 1.443 \exp\left(\frac{3.5645[eV]}{k_B T} \right) \left[\frac{cm^2}{s} \right] \tag{5}$$

$$D_i(B) = 0.797 \exp\left(-\frac{3.46[eV]}{k_B T} \right) \left[\frac{cm^2}{s} \right] \tag{6}$$

Diffusion coefficients during TED are 4 or 5 orders higher than those in thermal equilibrium as shown in Fig. **8**. If we assume that B diffusion is perfectly dominated by interstitial Si pair, and set $f_{Ieff}(B) = 1$, we then obtain that for In as

$$f_{Ieff}(In) = \frac{40}{270} \tag{7}$$

This means that the dominant diffusion mechanism is vacancy paring even for In, although we can observe TED. Since the enhanced factor for diffusion coefficient is around 4 orders, f_{Ieff} should be smaller than 10^{-4} to observe transient retarded diffusion instead of TED.

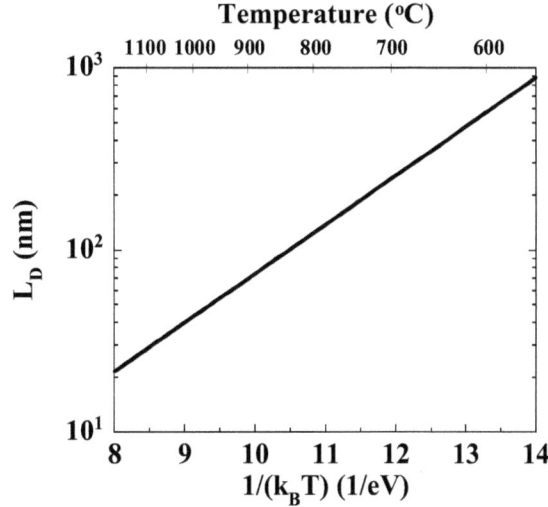

Figure 9: Dependence of diffusion length on temperature.

Let us evaluate the diffusion length associated with TED defined by

$$L_{enh} = 2\sqrt{D_{enh}t_{enh}} \tag{8}$$

Substituting the extracted values for D_{enh} and t_{enh}, we obtain.

$$
\begin{aligned}
L_{enh} &= 2\sqrt{0.797\exp\left(-\frac{3.46[eV]}{k_BT}\right)270\exp\left[\frac{0.487(eV)}{k_BT}\right]1.06\times10^{-18}\exp\left(\frac{4.21(eV)}{k_BT}\right)\text{sec}} \\
&= 1.51\times10^{-8}\exp\left[\frac{0.62(eV)}{k_BT}\right]cm \\
&= 0.151\exp\left[\frac{0.62(eV)}{k_BT}\right]nm
\end{aligned} \tag{9}
$$

It is noteworthy that the activation energy of L_{enh} is positive. This means that TED is more significant at low temperatures as shown in Fig. **9**.

Fig. **10** shows diffusion profiles after TED is completed, that is, the diffusion time is longer than t_{enh} as shown in Fig. **6**. TED is more significant at low temperatures as expected, which we can observe at temperatures of less than 750°C as shown in Fig. **10**(a). However, we cannot observe the difference in the low concentration region at temperatures higher than 800°C as shown in Fig. **10**(b). This is because TED starts and finishes during ramp up time period as we mentioned before, and the data are not fundamental ones related to the target temperature.

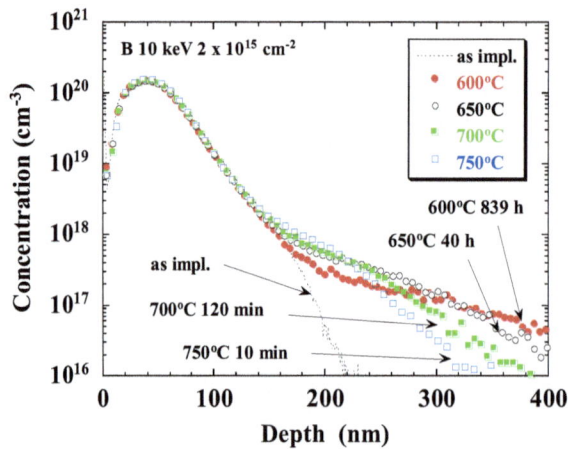

(a)

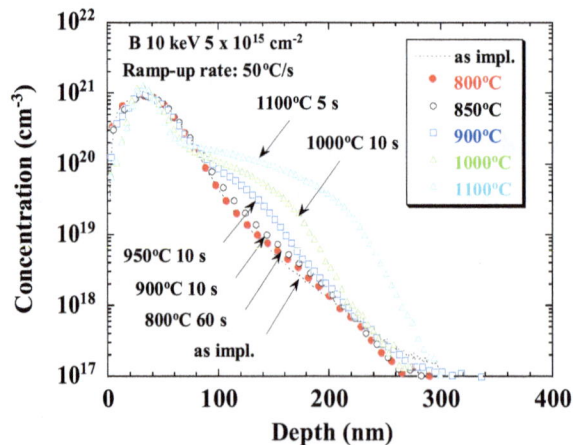

(b)

Figure 10: Diffusion profiles after TED. (a) Low temperature region. (b) High temperature region.

Inspecting above, we believe that the order of annealing process is expected to influence final junction depth. It is effective to anneal at high temperature first and then low temperature to obtain shallow junctions. In the practical VSLI process, this kind of situation exists. After the gate formation, extension ion implantation is done, and SiO_2 layer is then formed by chemical vapor deposition (CVD). The CVD process is typically at 600-700°C for few hours, and high temperature annealing is followed to activate the implanted ions. Fig. **11** shows the final diffusion profiles for different order of high and low temperature processes. The junction depth is deeper when we first anneal at low temperatures. Therefore, it is effective to insert high temperature process before extension process to obtain shallow junction.

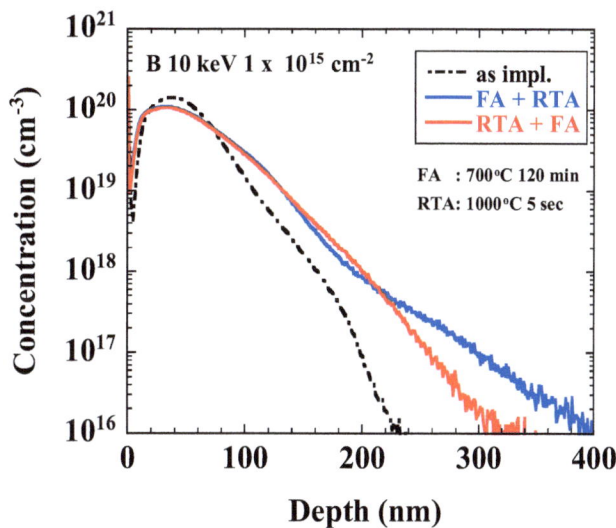

Figure 11: Difference of diffusion profiles between the order of FA and RTA.

DIFFUSION PROFILES USING B MARKER LAYER

TED is not important for As since its dominant diffusion mechanism is vacancy pair. However, As ion implantation induces excess point defects and the defects influence existing impurities which are sensitive to TED such as B. B marker layer in the deep region is frequently used to monitor the influence of the defects in the surface region induced by ion implantation. This system enables us to evaluate diffusion parameters focusing on the surface impurity profiles, and independently to evaluate the point defect parameters focusing on buried B marker layer. The systematic evaluations have been done, and the data were reproduced by three stream model simulation [11-13].

Fig. **12** shows dependence of B maker layer diffusion on B implantation dose. The annealing time of 60 sec at 800°C is almost the same as t_{enh}. The diffusion becomes significant with increasing dose. However, the dependence is not so significant since the dose changes two orders while the profile change is rather little.

Fig. **13** shows dependence of B maker layer diffusion on ion implantation energy. The diffusion becomes significant with increasing energy a little.

Fig. **14** shows dependence of B maker layer diffusion on As ion implantation dose. The diffusion increases with dose a little. The B marker layer diffusion is not so different from ones with B implantation.

Fig. **15** shows dependence of B maker layer diffusion on P ion implantation dose. The diffusion becomes significant with dose a little, and the diffusion is almost the same as B.

In the above simulation, the induced damage profile C_I is related to the ion implantation profile as

$$C_I = \begin{cases} f_D \Phi g(x) & for\ \Phi \leq \Phi_c \\ f_D \Phi_c g(x) & for\ \Phi > \Phi_c \end{cases} \tag{10}$$

where Φ is the dose, $g(x)$ is the ion implantation profile normalized with dose, f_D is a fitting parameter. Φ_c is the saturated dose where the TED saturates. The simulations above are done using the parameter set of Table **1**.

Table 1: f_D and Φ_c of B, As, and P

Impurity	B	As	P
f_D	2.0	2.0	1.8
$\Phi_c\left(\times 10^{14}\ cm^{-2}\right)$	1.1	1.3	1.5

Therefore, the induced damage does not have significant impurity dependence.

Inspecting above, the influence of induced damage to the buried layer is rather insensitive to ion implanted impurities. The energy and dose dependence are not so significant, either. These mean that the macroscopic features of TED do not have complex matrix to express, and simple treatment is possible, which we try in the next chapter.

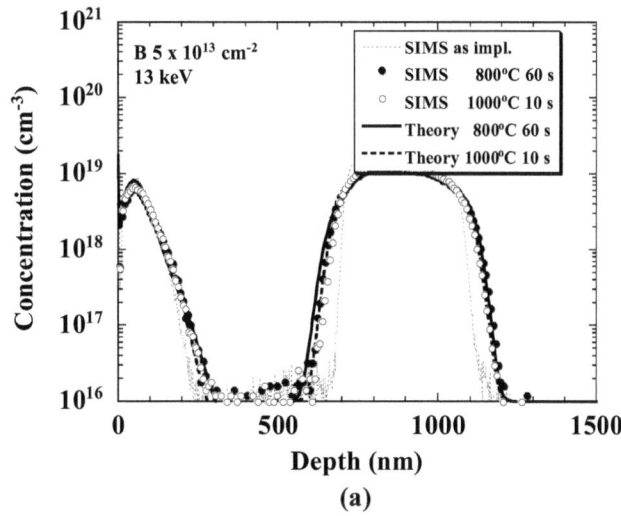

(a)

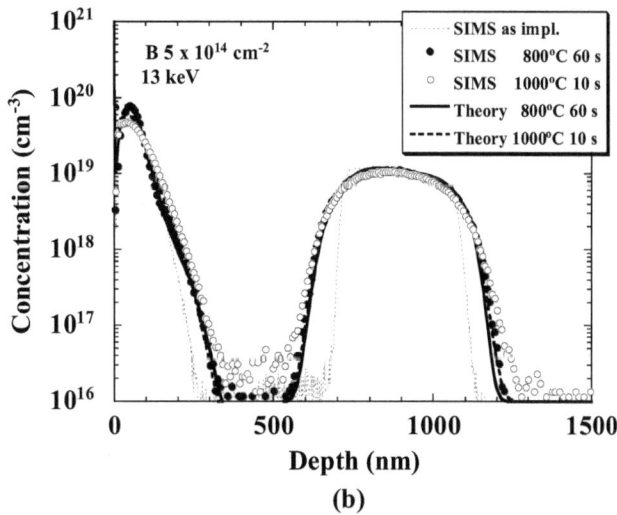

(b)

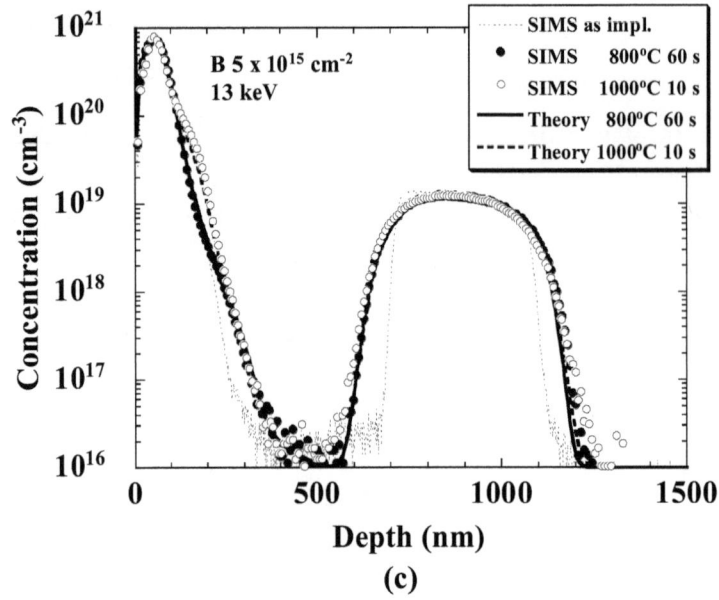

(c)

Figure 12: Dependence of TED profiles of B with buried B marker layer on dose. (a) $5 \times 10^{13} cm^{-2}$. (b) $5 \times 10^{14} cm^{-2}$. (c) $5 \times 10^{15} cm^{-2}$.

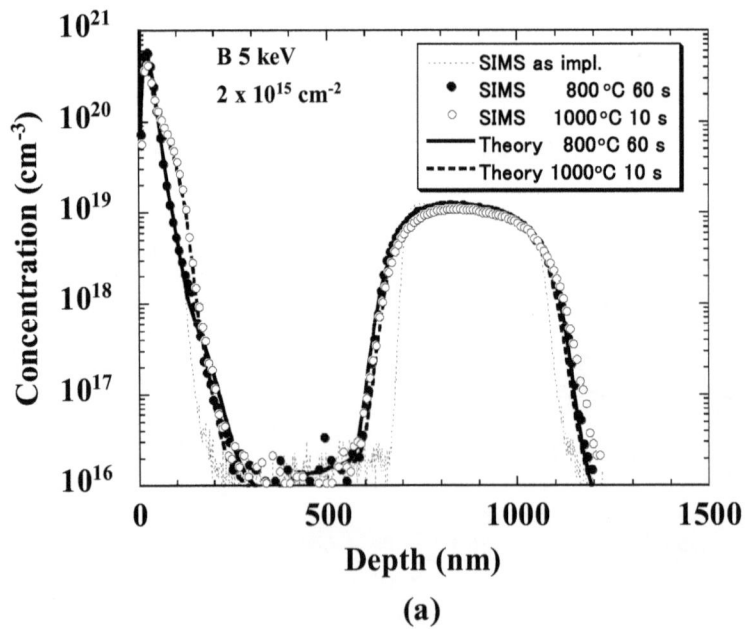

(a)

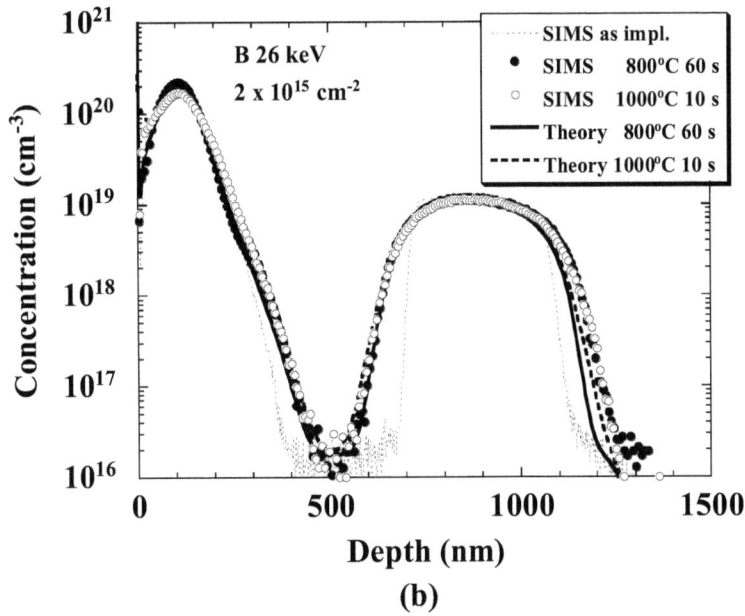

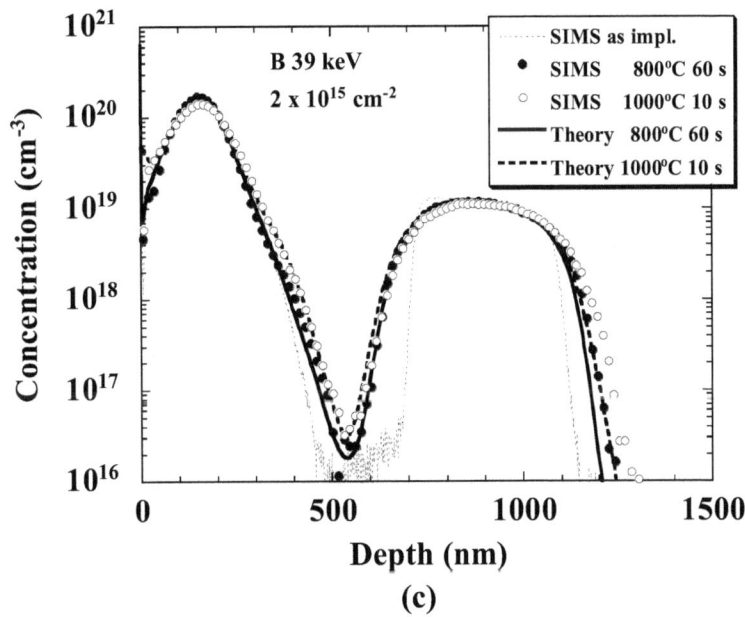

Figure 13: Dependence of TED profiles of B with buried B marker layer on energy. (a) 5 keV. (b) 26 keV. (c) 36 keV.

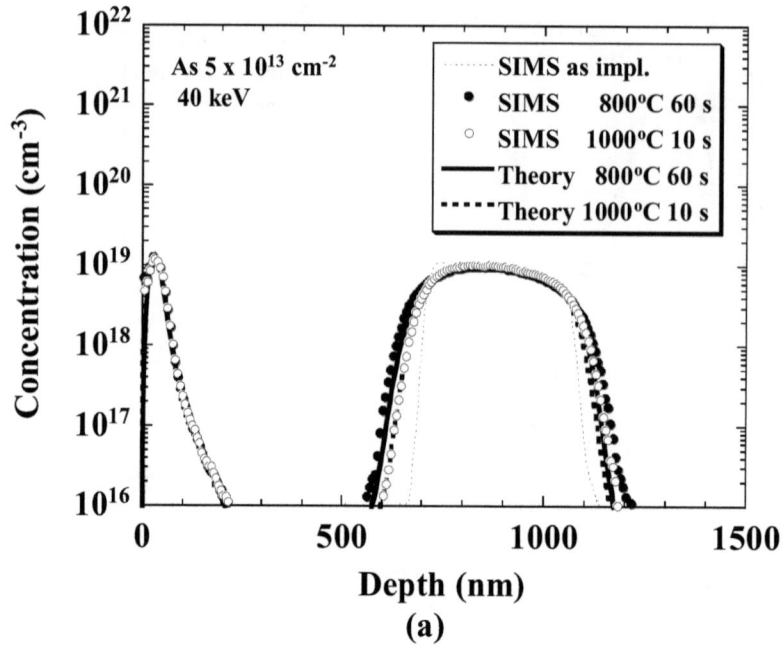

(a)

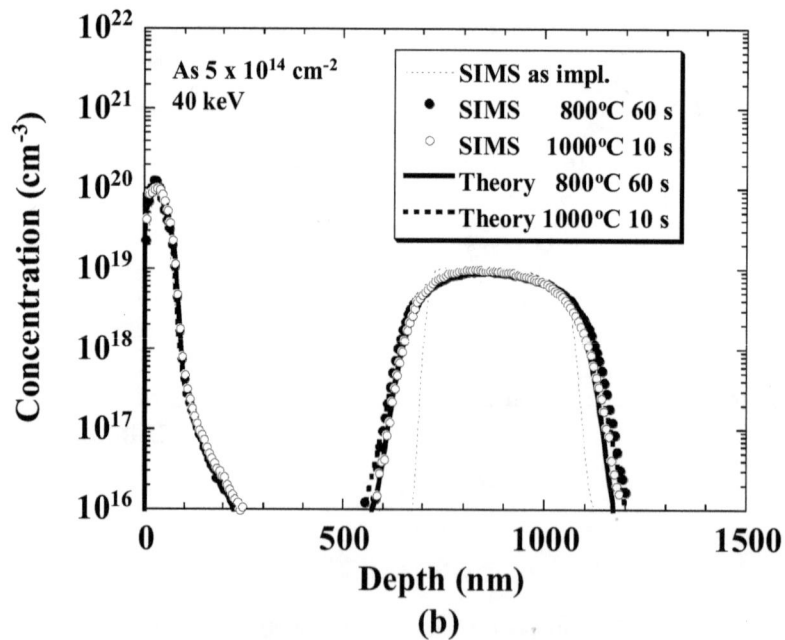

(b)

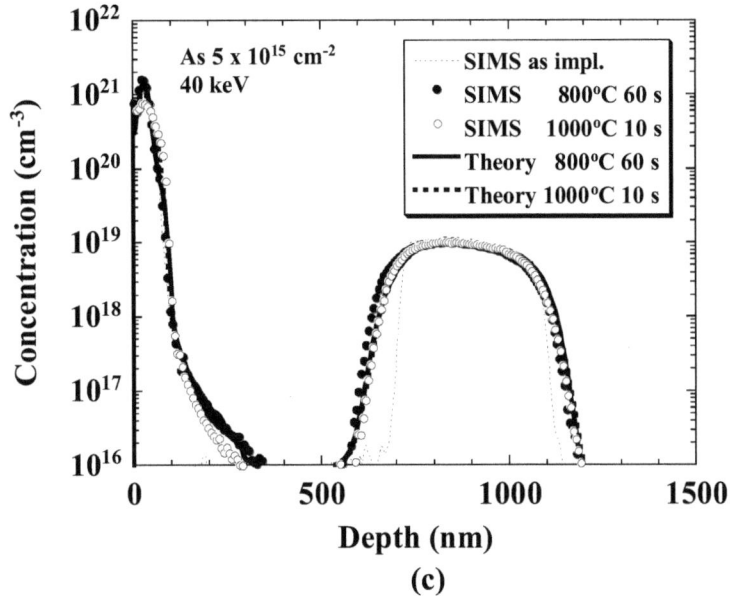

Figure 14: Dependence of TED profiles of As with buried B marker layer on As ion implantation dose. (a) $5 \times 10^{13} cm^{-2}$. (b) $5 \times 10^{14} cm^{-2}$. (c) $5 \times 10^{15} cm^{-2}$.

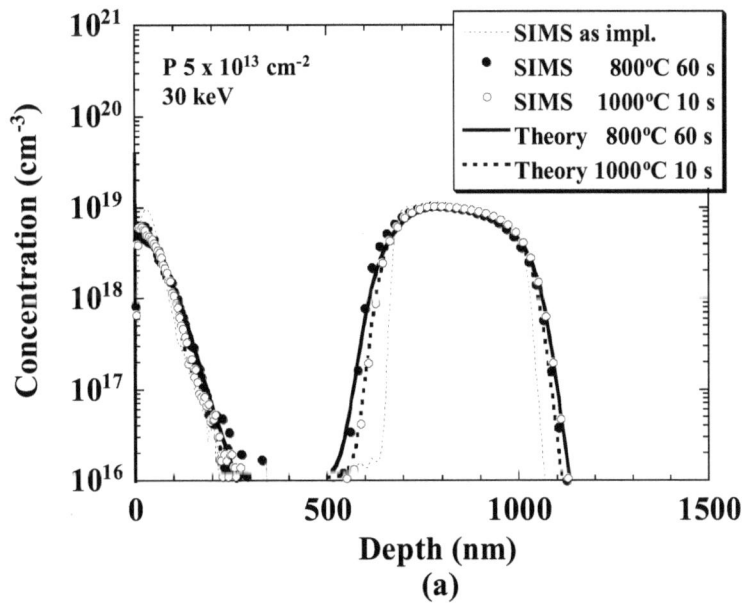

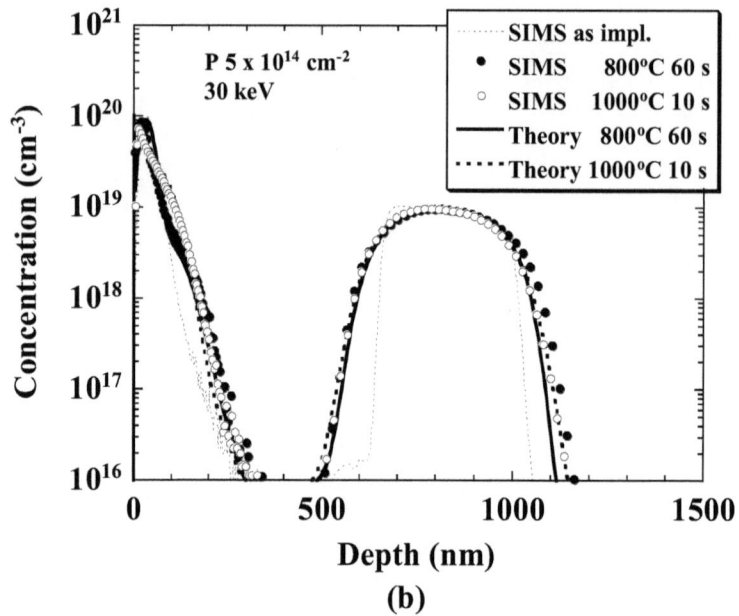

(b)

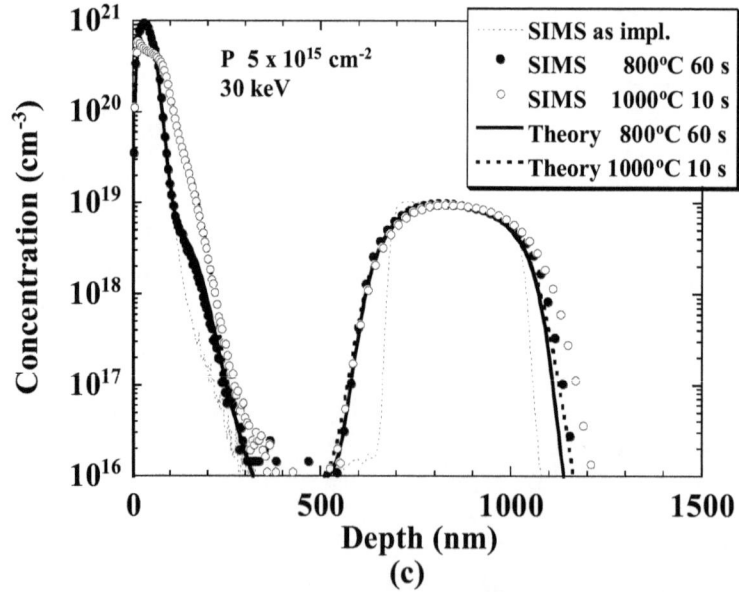

(c)

Figure 15: Dependence of TED profiles of P with buried B marker layer on P ion implantation dose. (a) $5 \times 10^{13} cm^{-2}$. (b) $5 \times 10^{14} cm^{-2}$. (c) $5 \times 10^{15} cm^{-2}$.

SUMMARY

Summarizing the above data, the macroscopic TED features are as follows.

The maximum diffusion concentration during TED is close to intrinsic carrier concentration.

The enhanced factor for diffusion coefficient TED is around of 3 and 4 orders.

The diffusion length related to TED is longer at lower annealing temperatures.

t_{enh} is rather insensitive to impurity.

The diffusion of marker layer is rather insensitive to kind of impurities that are ion implanted.

TED is more significant with increasing dose, and saturated when the dose reaches $10^{15} cm^{-2}$, an also more significant with increasing energy.

REFERENCES

[1]   T. Y. Tan, U. Goesele, "Point defects, diffusion processes and swirl defect formation in silicon," J. Appl. Phys., vol. 37, p. 1, 1985.

[2]   G. B. Bronner, J. D. Plummer, "Gettering of gold in silicon: A toll for understanding the properties of silicon interstitials," J. Appl. Phys., vol. 61, p. 5286, 1987.

[3]   H. Zimmermann, H. Rysse, "Gold and platinum diffusion: The key to the understanding of intrinsic point defect behavior in silicon," Appl. Phys. A, vol. 55, p. 121, 1992.

[4]   C. Boit, F. Lau, R. Sittig, "Gold diffusion in silicon by rapid optical annealing," Appl. Phys. A, vol. 50, p. 197, 1990.

[5]   F. F. Morehead, E. Rorris, R. R. O'Brien, R. F. Lever, J. P. Peng, G. R. Srinivasan, "Fast simulation of coupled defect-impurity diffusion in FINDPRO: Comparison with TSUPREM IV and other programs," J. Electrochem. Soc., vol. 138, p. 3789, 1991.

[6]   H. –J. Gossmann, C. S. Rafferty, H. S. Luftman, F. C. Unterwald, T. Boone, J. M. Poate, "Oxidation enhanced diffusion in Si B-doping superlattices and Si self-interstitial diffusivities," Appl. Phys. Lett., vol. 63, p. 639, 1993.

[7]   H. Bracht, N. A. Stolwijk, and H. Mehrer, "Properties of intrinsic point defects in silicon determined by zinc diffusion experiments under nonequilibrium conditions," Physical review B, vol. 52, No. 23, pp. 16542-16560, 1995.

[8]   A. Hoefler, T. Feudel, N. Strecker, W. Fichtner, K. Suzuki, Y. Kataoka, and N. Sasaki, "Modeling of boron diffusion after shallow implants using BF_2," 18th ECS meeting, pp.428-429, 1996.

[9]   A. Hoefler, T. Feudel, N. Strecker, W. Fichtner, K. Suzuki, N. Sasaki, and Y. Kataoka, "Modeling and simulation of spacial dependent transient diffusion after BF_2 implantation," 1996 SISPAD, pp. 13-14.

[10] K. Suzuki, M. Aoki, Y. Kataoka, N. Sasaki, A. Hoefler, T. Feudel, N. Strecker, and W. Fichtner," Analytical models for transient diffusion and activation of ion-implanted boron during rapid thermal annealing considering ramp-up period," International Electron Device Meeting, pp.799-802, 1996.

[11] K. Suzuki, T. Miyashita, Y. Tada, Alexander Hoefler, N. Strecker, and W. Fichtner, "Damage calibration concept and novel B cluster reaction model for B transient enhanced diffusion over thermal process range from $600°C$ (839 h) to $1100 °C$ (5s) with various Ion implantation doses and energies," International Electron Device Meeting, pp. 501-504, 1997.

[12] K. Suzuki, N. Strecker, and W. Fichtner, "Damage accumulation by arsenic ion implantation and its impact on transient enhanced diffusion of As and B," SISPAD, pp. 51-54, 1998.

[13] K. Suzuki, "Model for transient enhanced diffusion of ion-implanted boron, arsenic, and phosphorous over wide range of process conditions," Fujitsu Sci. Tech., vol. 39, pp. 138-149, 2003.

[14] A. Agarwal, T. E. Haynes, D. J. Eaglesham, H.-J. Gossmann, D. C. Jacobson, and J. M. Poate, and Y. E. Erokhin, "Interstitial defects in silicon from 1–5 keV Si1 ion implantation," Appl. Phys. Lett., vol. 70, p. 3332, 1997.

[15] S. Boninelli, N. Cherkashin, and A. Claverie, and F. Cristiano, "Evidences of an intermediate rodlike defect during the transformation of {113} defects into dislocation loops," Appl. Phys. Lett., vol. 89, 161904, 2006.

[16] P. Calvo, A. Claverie, N. Cherkashin, B. Colombeau, Y. Lamrani a, B. de Mauduit, and F. Cristiano, "Thermal evolution of {1 1 3}defects in silicon: transformation against dissolution," Nuc. Inst. Meth. Phys. Res. B, vol. 216, p. 173, 2004.

[17] B. Colombeau, N.E.B. Cowern, F. Cristiano, P. Calvo, Y. Lamrani,N. Cherkashin, E. Lampin, and A. Claverie, "Depth dependence of defect evolution and TED during annealing," Inst. Meth. Phys. Res. B, vol. 216, p. 90, 2004.

[18] Jinghong Li and Kevin S. Jones, "{311} defects in silicon: The source of the loops," Appl. Phys. Lett., vol. 73, p. 3478, 1998.

[19] K. Suzuki, H. Tashiro, and T. Aoyama, "Diffusion coefficient of indium in Si substrates and analytical redistribution model," Solid-State Electronics, vol. 43, pp. 27-31, 1999.

[20] S. M. Sze, *VLSI Technology*, McGRAW-HILL, 1988.

Send Orders for Reprints to reprints@benthamscience.net
Ion Implantation and Activation, Vol. 3, 2013, 83-100 **83**

CHAPTER 4

Simple Treatment of Transient Enhanced Diffusion

Abstract: A two-stream model which describes impurity flux and point defect flux independently has been proposed. Equation for point defect flux gives macroscopic information about the impurity flux equation, such as enhanced diffusion coefficient and time duration of TED. The model also accommodates the ramp up process where enhanced diffusion and time duration parameters become variable. This simple model can be controlled and tuned simply and gives us prominent features of TED.

Keywords: Ion implantation, transient enhanced diffusion, two-stream model, point defect, diffusion coefficient, pairing, damage, annealing, electric field, interstitial Si, vacancy, intrinsic carrier concentration, solid solubility, dose, amorphous/crystal interface, (311) defect, diffusion length, dislocation loop.

INTRODUCTION

Transient enhanced diffusion (TED) can be described by treating pairing of impurity and point defects with three and five stream models, as described in chapter 2. The associated pairing and reaction rate parameters should be tuned well for robust simulation, and intensive work on these subjects has been done.

In this chapter, we treat TED empirically and describe the following content.

(a) The amount of damage induced by ion implantation.

(b) Location of the damage region with a delta function form.

(c) Enhanced diffusion coefficient.

(d) Maximum diffusion concentration during TED.

(e) Time duration of TED.

These parameters are related to each other and they should be simply determined according to vast matrix of ion implantation and annealing conditions.

THEORETICAL FRAMEWORK

Here, we explain theoretical framework of the empirical model.

The diffusion equation we use is given by

$$\frac{\partial N(x)}{\partial t} = \frac{\partial \left[K_{ele} K_{pdef} D \dfrac{\partial N(x)}{\partial x} \right]}{\partial x} \tag{1}$$

where N is the impurity concentration, and D is the diffusion coefficient given by

$$D = D_i^{(x)} + \left(\frac{p}{n_i} \right) D_i^{(p)} + \left(\frac{n}{n_i} \right) D_i^{(m)} + \left(\frac{n}{n_i} \right)^2 D_i^{(mm)} \tag{2}$$

p is the whole concentration, n is the electron concentration, and n_i is the intrinsic carrier concentration. K_{ele} expresses the electric field influence and is given by

$$K_{ele} \equiv \left| 1 + \frac{1}{\sqrt{1 + \left(\dfrac{4n_i}{N} \right)^2}} \right| \tag{3}$$

K_{pdef} expresses the point defect influence and is given by

$$K_{pdef} \equiv \frac{[I]}{[I]^*} f_{Ieff} + \frac{[V]}{[V]^*} \left(1 - f_{Ieff} \right) \tag{4}$$

where $\lfloor I^* \rfloor$ and $\lfloor V^* \rfloor$ are concentrations of interstitial Si and vacancy in thermal equilibrium, respectively, and those without symbol of * are concentrations in a general case.

Setting $[I] = \lfloor I^* \rfloor$ and $[V] = \lfloor V^* \rfloor$, we obtain the diffusion coefficient in thermal equilibrium as

$$D^* = K_{ele}D = \left| 1 + \frac{1}{\sqrt{1+\left(\frac{4n_i}{N}\right)^2}} \right| \left[D_i^{(x)} + \left(\frac{p}{n_i}\right)D_i^{(p)} + \left(\frac{n}{n_i}\right)D_i^{(m)} + \left(\frac{n}{n_i}\right)^2 D_i^{(mm)} \right] \qquad (5)$$

The impurity concentration profiles in thermal equilibrium can be evaluated with the following diffusion equation.

$$\frac{\partial N(x)}{\partial t} = \frac{\partial \left[D^* \dfrac{\partial N(x)}{\partial x} \right]}{\partial x} \qquad (6)$$

It is noted that the maximum diffusion concentration is limited by solid solubility $N_{diffMax}$, that is,

$$N_{diffMax} = N_{sol} \qquad (7)$$

The diffusion coefficient during TED is different from those in thermal equilibrium, and it can be regarded as constant as shown in chapter 3. The diffusion equation is then given by

$$\frac{\partial N(x)}{\partial t} = \frac{\partial \left[D_{enh} \dfrac{\partial N(x)}{\partial x} \right]}{\partial x} \qquad (8)$$

where

$$D_{enh} = K_{ele}K_{pdef}D \qquad (9)$$

The maximum diffusion concentration is the one specially related to TED as N_{TEDMax}

$$N_{diffMax} = N_{TEDMax} \qquad (10)$$

Since the N_{TEDMax} is low and close to intrinsic carrier concentration, Eq. 8 may be reduced to

$$\frac{\partial N(x)}{\partial t} = D_{enh} \frac{\partial^2 N(x)}{\partial x^2} \tag{11}$$

and

$$D_{enh} = K_{pdef} D_i \tag{12}$$

where

$$D_i = D_i^{(x)} + D_i^{(p)} + D_i^{(m)} + D_i^{(mm)} \tag{13}$$

DAMAGE INDUCED BY ION IMPLANTATION

The damage should be related to ion implantation dose Φ. It is proportional to the implantation dose in low dose region, and damage overlaps with each other with increasing implantation dose, and then it saturates. The dose of the damage can be expressed by

$$Q_I = \begin{cases} r_I \Phi & for \ r_I \Phi \le Q_{Isat} \\ Q_{Isat} & for \ r_I \Phi > Q_{Isat} \end{cases} \tag{14}$$

where r_I is a proportional constant, and it should be determined depending onion implanted impurities. Q_{Isat} is the saturated damage dose.

The location of the damage region is related to the ion implantation profiles. The damage profile is simply expressed as the ion implantation concentration profile multiplied by a factor. The damage is set to be zero when the continuous amorphous layer is formed [2, 3].

Here, we treat the damage profiles with a delta function form located at d_I. We set it to R_p when the continuous amorphous layer is not formed, and set it at the amorphous/crystal (a/c) interface when continuous amorphous layer is formed as shown in Fig. **1**. We modify the amorphous layer thickness model [1, 2] and propose a model for d_I as

$$d_I = \begin{cases} R_p & \text{for } \Phi \leq 2\Phi_{a/c} \\ R_p + \sqrt{2}\Delta R_p erfc^{-1}\left(\dfrac{2\Phi_{a/c}}{\Phi}\right) & \text{for } \Phi > 2\Phi_{a/c} \end{cases} \tag{15}$$

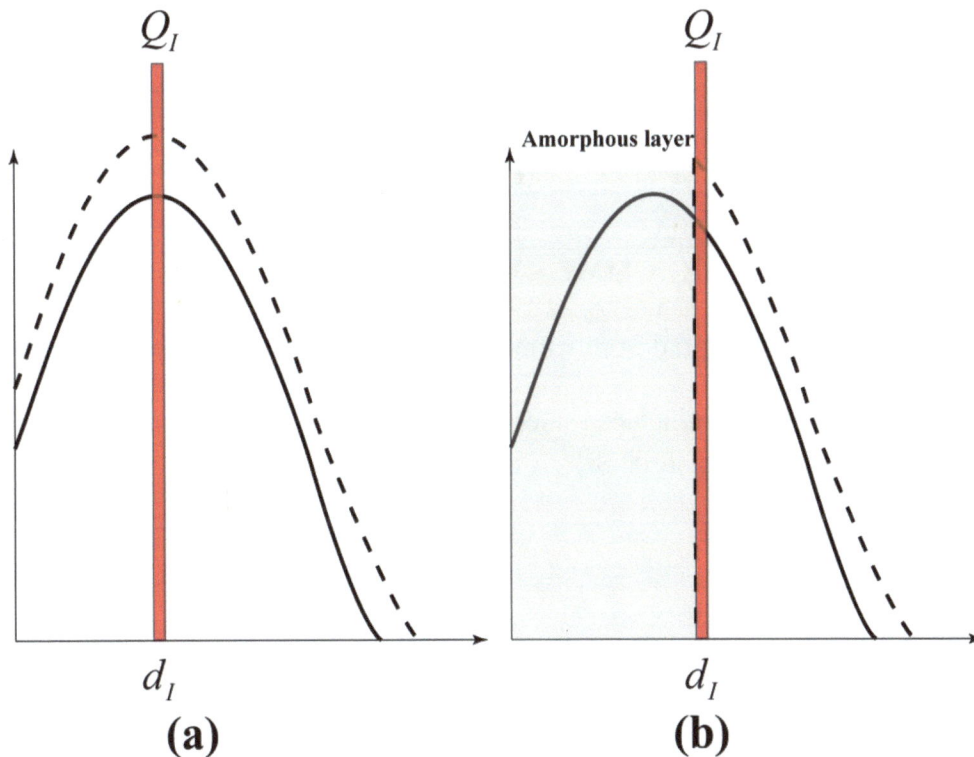

Figure 1: Schematic location of interstitial Si cluster. (a) Before forming amorphous layer. (b) After forming amorphous layer.

ENHANCED DIFFUSION COEFFICIENT

The induced point defects form stable (311) defect clusters, and the clusters release mobile point defects [3-8]. We assume that all of the induced point defects form a cluster, and it plays a role of constant concentration diffusion source, and we denote the concentration as   . The enhanced diffusion coefficient is multiplied by a factor I_{sol}/I^*, and the vacancy concentration V is given by

$$I_{sol}V = I^*V^*$$
(16)

we then obtain

$$V = \frac{I^*}{I_{sol}}V^*$$
(17)

We do not consider the formation of a vacancy cluster.

The enhanced factor associated with TED is then given by

$$\frac{I}{I^*}f_{Ieff} + \frac{V}{V^*}\left(1 - f_{Ieff}\right) = \frac{I_{sol}}{I^*}f_{Ieff} + \frac{I^*}{I_{sol}}\left(1 - f_{Ieff}\right)$$
(18)

ENHANCED DIFFUSION TIME

We show various interstitial Si flux models, from a simple one to the general one.

We first assume that the induced interstitial Si vanishes only at the surface. The flux associated with interstitial Si f_I is then given by

$$f_I = D_I \frac{I_{sol}}{d_I}$$
(19)

where D_I is the diffusion coefficient of interstitial Si (Fig. **2**). We assume that the flux is invariable during TED. When time passed for a time period of t_{enh}, the integral of the flux reaches the total amount of induced point defect of Q_I, and TED then finishes. That is,

$$D_I \frac{I_{sol}}{d_I} t_{enh} = Q_I$$
(20)

Therefore, we obtain

$$t_{enh} = \frac{d_I Q_I}{D_I I_{sol}}$$
(21)

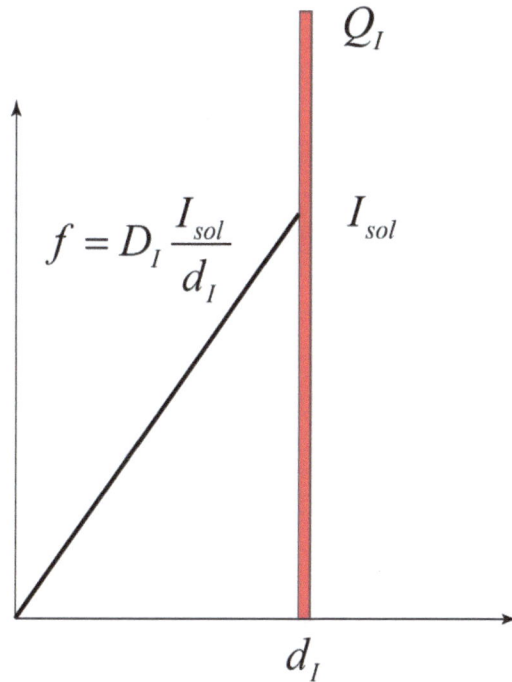

$$Q_I$$

$$f = D_I \frac{I_{sol}}{d_I}$$

$$I_{sol}$$

$$d_I$$

Figure 2: Intestinal Si flux towards the surface.

Assuming f_{Ieff} is 1, the diffusion length L_{Denh} associated with TED is then expressed by

$$
\begin{aligned}
L_{Denh} &= 2\sqrt{D_{enh}t_{enh}} \\
&= 2\sqrt{\frac{I_{sol}}{I^*} D \frac{d_I Q_I}{D_I I_{sol}}} \\
&= 2\sqrt{D \frac{d_I Q_I}{D_I I^*}}
\end{aligned}
\tag{22}
$$

Note that the diffusion length does not depend on   . It depends on the product of D_I and   , which is thought to be well established value. The ambiguous parameter in Eq. 22 is Q_I.

We assume that the surface is a perfect sink of the interstitial Si in Eq. 19. We can extend the model to treat the surface in general introducing reaction rate h. We assume

$$D_I \frac{I_{sol} - I_s}{d_I} = hI_s \tag{23}$$

where I_s is the interstitial concentration at the surface. We then obtain I_s as

$$I_S = \frac{\dfrac{D_I}{d_I}}{\dfrac{D_I}{d_I} + h} I_{sol} \tag{24}$$

The flux is then given by

$$f_I = D_I \frac{I_{sol} - \dfrac{\dfrac{D_I}{d_I}}{\dfrac{D_I}{d_I} + h} I_{sol}}{d_I} \tag{25}$$

$$= \frac{1}{\dfrac{D_I}{d_I h} + 1} D_I \frac{I_{sol}}{d_I}$$

Therefore, the flux in the limiting cases are expressed by

$$f_I = \begin{cases} D_I \dfrac{I_{sol}}{d_I} & \text{for } h \to \infty \\ hI_{sol} & \text{for } h \to 0 \end{cases} \tag{26}$$

The recombination in the bulk region is neglected in Eq. 25. Let us consider how we include this. The diffusion equation for interstitial Si in bulk Si region can be expressed by

$$\frac{\partial I}{\partial t} = D_I \frac{\partial^2 I}{\partial x^2} - \frac{I}{\tau_I} \tag{27}$$

where τ_I is the recombination time. We assume that the steady state is hold during TED, and Eq. 27 is reduced to

$$D_I \frac{\partial^2 I}{\partial x^2} - \frac{I}{\tau_I} = 0 \qquad (28)$$

The interstitial Si concentration profile in the deeper region is then solved as

$$I(x) = I_{sol} \exp\left(-\frac{x}{L_{DI}}\right) \qquad (29)$$

where L_{DI} is the diffusion length of interstitial Si defined by

$$L_{DI} = \sqrt{D_I \tau_I} \qquad (30)$$

The origin is set at the location of interstitial cluster, that is, d_I. The flux at $x = 0$ is then given by

$$\begin{aligned}
f_I &= -D_I \left.\frac{\partial I}{\partial x}\right|_{x=0} \\
&= \frac{D_I I_{sol}}{L_{DI}}
\end{aligned} \qquad (31)$$

Therefore, t_{enh} is then given by

$$t_{enh} = \frac{Q_I}{\dfrac{1}{\dfrac{D_I}{d_I h} + 1} D_I \dfrac{I_{sol}}{d_I} + \dfrac{D_I I_{sol}}{L_{DI}}} \qquad (32)$$

By comparing with experimental data, we should select a model.

In the above discussion, we assume that interstitial Si vanishes through the recombination. It is also proposed that point defects form dislocation loops, and the dislocation loops are stable and do not release interstitial Si. We may be able to empirically express this effect with τ_I.

MAXIMUM DIFFUSION CONCENTRATION DURING TED

The maximum diffusion concentration during TED is constant and close to intrinsic carrier concentration as shown in chapter 3. We denote the concentration as N_{TEDMax}. This should be explained more physically.

TREATMENT OF RAMP UP PROCESS

TED finishes in quite a short time at high temperatures. We can set parameters related to the variable temperatures. We should monitor whether it is on the way of TED or after finishing TED.

We assume that the initial absolute temperature T_0 with ramp up rate of r_{rampU}. The temperature at time t, $T(t)$, is then given by

$$T(t) = T_0 + r_{rampU} t \tag{33}$$

We assume that TED finishes at the temperature of T_f and corresponding TED time is Δt_f. The effective time at the temperature T is related as

$$\frac{\Delta t}{t_{enh}(T(t))} = \frac{\Delta t_f}{t_{enh}(T_f)} \tag{34}$$

Therefore, we obtain

$$\Delta t_f = \frac{t_{enh}(T_f) \Delta t}{t_{enh}(T(t))} \tag{35}$$

Integrating Eq. 35, we obtain

$$t_{enh}(T_f) = \int_0^{t_f} \frac{t_{enh}(T_f)}{t_{enh}(T(t))} dt \tag{36}$$

That is, we can evaluate the finishing TED time with

$$1 = \int_0^{t_f} \frac{1}{t_{enh}(T_0 + r_{rampU} t)} dt \tag{37}$$

The temperature where TED finishes is

$$T_f = T_0 + r_{rampU}t_f \tag{38}$$

If TED does not finish during TED, we evaluate effective TED time during ramp up process, and the effective time for TED during target temperature is evaluated by

$$t_{eff} = \int_0^{t_{ramp}} \frac{1}{t_{enh}\left(T_0 + r_{rampU}t\right)}dt \times t_{enh}\left(T_0 + r_{rampU}t_{ramp}\right) \tag{39}$$

The TED time related to the isothermal process is

$$t_{enh}\left(T_0 + r_{rampU}t_{ramp}\right) - t_{eff} \tag{40}$$

After this time period, we solve the diffusion equation for thermal equilibrium.

PARAMETER VALUES

Here, we show the model parameter values. Most of which are not well established, and we should continue to study these values.

D_I and

The product of D_I and   is reported to be [9]

$$D_I I^* = 4.57 \times 10^{25} \exp\left[-\frac{4.84eV}{k_B T}\right] \quad cm^{-1}s^{-1} \tag{41}$$

This is believed to be well established value. The activation energy of impurity diffusion coefficients are also well established values, and all of them are less than 4.84 eV. Therefore, the activation energy of TED diffusion length is robustly predicted to be negative. This means that we can robustly insist on significant TED at low temperatures.

Various values for D_I were reported [10-15] as

$$D_I \left(\frac{cm^2}{s} \right) = \begin{cases} 600 \exp\left[-\dfrac{2.44 eV}{k_B T} \right] & Bronner\,[10] \\[3mm] 2.58 \times 10^{-2} \exp\left[-\dfrac{0.965 eV}{k_B T} \right] & Zimmermann\,[11] \\[3mm] 1.03 \times 10^6 \exp\left[-\dfrac{3.22 eV}{k_B T} \right] & Boit\,[12] \\[3mm] 3.6 \times 10^{-4} \exp\left[-\dfrac{1.58 eV}{k_B T} \right] & Morehead\,[13] \\[3mm] 1.0 \times 10^{-2} \exp\left[-\dfrac{3.1 eV}{k_B T} \right] & Gossmann\,[14] \\[3mm] 51 \exp\left[-\dfrac{1.77 eV}{k_B T} \right] & Bracht\,[15] \end{cases} \tag{42}$$

which is wide scattered as shown in Fig. **3**, and it is still under discussion.

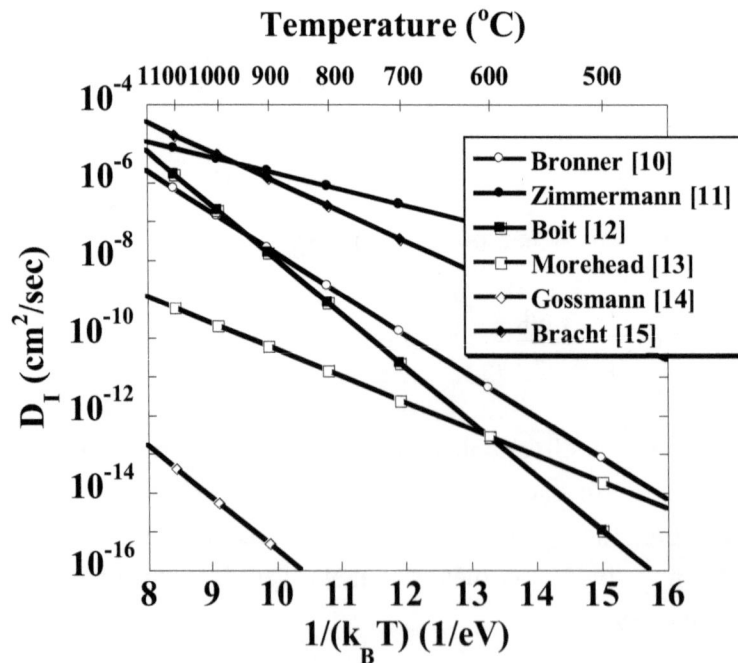

Figure 3: Various reported D_I.

From the standpoint of diffusion potential, Bronner, Bracht, and Morehead are plausible as shown in chapter 1, and we adopt here the Bronner model. We then obtain the interstitial Si concentration in thermal equilibrium as

$$I^* = 7.5 \times 10^{22} \exp\left[-\frac{2.4eV}{k_B T} \right] \quad cm^{-3} \qquad (43)$$

which is shown in Fig. **4**.

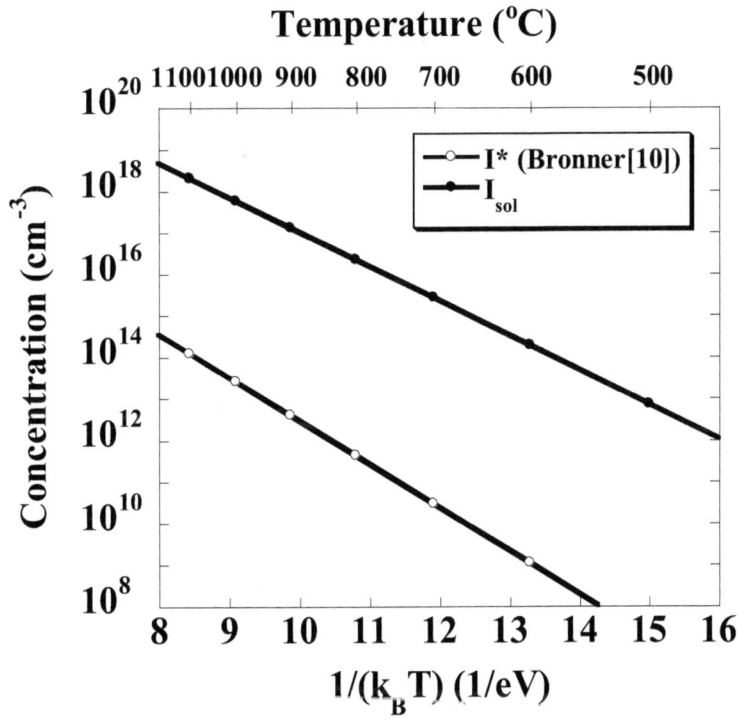

Figure 4: Dependence of I^* and I_{sol} on temperature.

Diffusion coefficient during TED D_{enh} is given by

$$D_{enh} \approx f_{Ieff} \frac{I_{sol}}{I^*} D \qquad (44)$$

We obtained D_{enh} in the previous section and assuming $f_{Ieff} \approx 1$ for B, we can evaluate   as

$$I_{sol} = \frac{D_{enh}}{D} I^*$$ (45)

and obtain

$$I_{sol} = 2.03 \times 10^{25} \exp\left(-\frac{1.913(eV)}{k_B T}\right)(cm^{-3})$$ (46)

which is shown in Fig. **4**. I_{sol} is higher than I^* by about 4 and 5 orders over wide temperature range.

Further, t_{enh} is also extracted as

$$t_{enh} = 2.61 \times 10^{-18} \exp\left(\frac{4.0[eV]}{k_B T}\right) \sec$$ (47)

where B ion implantation condition is 10 keV with a dose of 5 x 10^{15} cm^{-2}. With this ion implantation condition, continuous amorphous layer is not formed, and hence.

$$d_I = R_p = 38.4 nm$$ (48)

Substituting this into Eq. 18, we obtain

$$Q_I = 5 \times 10^{14} cm^{-2}$$ (49)

This induced damage is that in high dose region, and can be regarded as Q_{Isat}. We adopt $r_I = 2$.

Similar analysis is applied to In. t_{enh} is the same as that for B. $d_I = 42.0\,nm$ and hence Q_I is then evaluated as

$$Q_I = 6.4 \times 10^{14} cm^{-2}$$ (50)

We should use $r_I \approx 10$, which is much higher than that for B. In is trapped to the interstitial clusters and it may play a role of higher dose of interstitial Si in the cluster. The detail is not clear.

We regard that Q_{Isat} is rather insensitive to impurity and set its default value of $Q_I = 5 \times 10^{14} cm^{-2}$.

N_{enhMax}

The maximum diffusion concentration is close to intrinsic carrier concentration n_i for both B and In ion implantation. The origin is not clear, but we set that

$$N_{TEDMax} = n_i \tag{51}$$

where n_i is [16]

$$n_i = 3.87 \times 10^{16} T^{\frac{3}{2}} \exp\left[-\frac{0.605(eV)}{k_B T}\right] cm^{-3}$$

$$= 2.01 \times 10^{20} \left(\frac{T}{300}\right)^{\frac{3}{2}} \exp\left[-\frac{0.605(eV)}{k_B T}\right] cm^{-3} \tag{52}$$

SUMMARY

Summarizing above, we use the parameters of the empirical TED model as follows.

$$I^* = 7.5 \times 10^{22} \exp\left[-\frac{2.4eV}{k_B T}\right] cm^{-3} \tag{53}$$

$$I_{sol} - 2.03 \times 10^{25} \exp\left(-\frac{1.913(eV)}{k_B T}\right)(cm^{-3}) \tag{54}$$

$$D_I = 600 \exp\left[-\frac{2.44eV}{k_B T}\right] cm^2/s \tag{55}$$

$$N_{TEDMax} = n_i \tag{56}$$

$$d_I = \begin{cases} R_p & for\ \Phi \leq 2\Phi_{\%_c} \\ R_p + \sqrt{2}\Delta R_p\, erfc^{-1}\left(\dfrac{2\Phi_{\%_c}}{\Phi}\right) & for\ \Phi > 2\Phi_{\%_c} \end{cases} \tag{57}$$

$$Q_I = \begin{cases} r_I\Phi & for\ r_I\Phi \leq Q_{Isat} \\ Q_{Isat} & for\ r_I\Phi > Q_{Isat} \end{cases} \tag{58}$$

$$t_{enh} = \frac{d_I Q_I}{D_I I_{sol}} \tag{59}$$

In more general, we use

$$t_{enh} = \frac{Q_I}{\dfrac{1}{\dfrac{D_I}{d_I h}+1}D_I\dfrac{I_{sol}}{d_I} + \dfrac{D_I I_{sol}}{L_{DI}}} \tag{60}$$

Table **1** summarized the parameter values for transient enhanced diffusion for various impurities.

Table 1: Parameter values for transient enhanced diffusion for impurities

Impurity	B	In	P	As	Sb	unknown
f_{Ieff}	1	0.15	0.1	0	0	0
r_I	2	10	2	2	2	2
$Q_{Isat}\ (cm^{-2})$	5×10^{14}	5×10^{14}	5×10^{14}	5×10^{14}	5×10^{14}	5×10^{14}

Using above, the diffusion during TED is described as

$$\frac{\partial N(x)}{\partial t} = \frac{\partial\left[D_{enh}\dfrac{\partial N(x)}{\partial x}\right]}{\partial x} \tag{61}$$

$$D_{enh} = \left| \frac{I_{sol}}{I^*} f_{Ieff} + \frac{I^*}{I_{sol}} \left(1 - f_{Ieff} \right) \right| D^*$$ (62)

$$N_{diffMax} = N_{TEDMax}$$ (63)

After TED finishes, the diffusion is described as

$$\frac{\partial N(x)}{\partial t} = \frac{\partial \left[D^* \frac{\partial N(x)}{\partial x} \right]}{\partial x}$$ (64)

$$N_{diffMax} = N_{sol}$$ (65)

The definition whether TED finishes or not is evaluated as follows.

The time when TED finishes is evaluated by

$$1 = \int_0^{t_f} \frac{1}{t_{enh} \left(T_0 + r_{rampU} t \right)} dt$$ (66)

The corresponding temperature is given by

$$T_f = T_0 + r_{rampU} t_f$$ (67)

REFERENCES

[1]   K. Suzuki, K. Kawamura, Y. Kikuchi, and Y. Kataoka, "Compact model for amorphous layer thickness formed by ion implantation over wide ion implantationconditions," IEEE Trans. Electron Devices, vol. ED-53, NO. 5, pp. 1186-1192, 2006.

[2]   K. Suzuki, Y. Tada, Y. Kataoka, K. Kawamura, and T. Nagayama, "Analytical model of amorphous-layer thickness formed by high-tilt-angle As ion implantation," IEEE Trans. Electron Devices, vol. ED-55, NO. 4, pp. 1080-1084, 2008.

[3]   A. Agarwal, T. E. Haynes, D. J. Eaglesham, H.-J. Gossmann, D. C. Jacobson, and J. M. Poate, and Yu. E. Erokhin, "Interstitial defects in silicon from 1–5 keV Si1 ion implantation," Appl. Phys. Lett., vol. 70, p. 3332, 1997.

[4]   S. Boninelli, N. Cherkashin, and A. Claverie, and F. Cristiano, "Evidences of an intermediate rodlike defect during the transformationof {113} defects into dislocation loops," Appl. Phys. Lett., vol. 89, 161904, 2006.

[5]   P. Calvo, A. Claverie, N. Cherkashin, B. Colombeau,Y. Lamrani a, B. de Mauduit, and F. Cristiano, "Thermal evolution of {1 1 3}defects in silicon:transformation against dissolution," Nuc. Inst. Meth. Phys. Res. B, vol. 216, p. 173, 2004.

[6]   B. Colombeau, N.E.B. Cowern, F. Cristiano, P. Calvo, Y. Lamrani,N. Cherkashin, E. Lampin, and A. Claverie, "Depth dependence of defect evolution and TEDduring annealing," Inst. Meth. Phys. Res. B, vol. 216, p. 90, 2004.

[7]   Jinghong Li and Kevin S. Jones, "{311} defects in silicon: The source of the loops," Appl. Phys. Lett., vol. 73, p. 3478, 1998.

[8]   Hugo Saleh, Mark E. Lawa, Sushil Bharatan, Kevin S. Jones, and Viswanath Krishnamoorthy, and Temel Buyuklimanli, "Energy dependence of transient enhanced diffusion and defect kinetics," Appl. Phys. Lett., vol. 77, p. 112, 2000.

[9]   T. Y. Tan, U. Goesele," Point defects, diffusion processes and swirl defect formation in silicon," J. Appl. Phys., vol. 37, p. 1, 1985.

[10]   G. B. Bronner, J. D. Plummer, "Gettering of gold in silicon: A toll for understanding the properties of silicon interstitials," J. Appl. Phys., vol. 61, p. 5286, 1987.

[11]   H. Zimmermann, H. Ryssel, "Gold and platinum diffusion: The key to the understanding of intrinsic point defect behavior in silicon," Appl. Phys. A, vol. 55, p. 121, 1992.

[12]   C. Boit, F. Lau, R. Sittig, "Gold diffusion in silicon by rapid optical annealing," Appl. Phys. A, vol. 50, p. 197, 1990.

[13]   F. F. Morehead, E. Rorris, R. R. O'Brien, R. F. Lever, J. P. Peng, G. R. Srinivasan, "Fast simulation of coupled defect-impurity diffusion in FINDPRO: Comparison with TSUPREM IV and other programs," J. Electrochem. Soc., vol. 138, p. 3789, 1991.

[14]   H. –J. Gossmann, C. S. Rafferty, H. S. Luftman, F. C. Unterwald, T. Boone, J. M. Poate, "Oxidation enhanced diffusion in Si B-doping superlattices and Si self-interstitial diffusivities," Appl. Phys. Lett., vol. 63, p. 639, 1993.

[15]   H. Bracht, N. A. Stolwijk, and H. Mehrer, "Propertieds of intrinsic point defects in silicon determined by zinc diffusion experiments under nonequilibrium conditions," Physical review B, vol. 52, No. 23, pp. 16542-16560, 1995.

[16]   F. J. Morin and J. P. Maita, "Electrical properties of silicon containing arsenic and boron," Physical Review, Vol. 96, pp. 28-35, 1954.

Send Orders for Reprints to reprints@benthamscience.net
Ion Implantation and Activation, Vol. 3, 2013, 101-119 101

CHAPTER 5

Thermal Oxidation

Abstract: Oxidation model is derived by considering diffusion fluxes in gas phase atmosphere, growing oxide layer, and reaction of oxidant and substrate Si atoms at SiO$_2$/Si interface. This model gives simple time dependence of growing SiO$_2$ layer thickness. The impurities in the Si substrate redistribute during the oxidation. We treat the redistributed profile as moving boundary one, and derive the corresponding model. The model well predicts B depletion at the SiO$_2$/Si interface.

Keywords: Oxidation, segregation, SiO$_2$, VLSI, MOS, chemical reaction, redistribution, surface, mass-transfer constant, Henry's law, ideal gas law, SiO$_2$/Si interface, linear dependence, parabolic dependence, Massoud model.

INTRODUCTION

SiO$_2$ layer plays an important role in VLSI processing. It isolates each device, and is also used for controlling charge concentration in MOS devices. Therefore, the growth of SiO$_2$ layer is a basic feature of the planar technology. The precise control of the SiO$_2$ growth thickness and the understanding of the oxide growth are important.

SiO$_2$ layers on Si substrate can be formed by thermal oxidation of Si through chemical reaction of

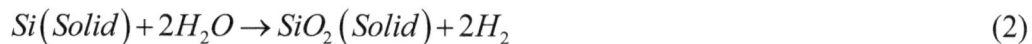

$$Si\left(Solid\right)+O_2 \rightarrow SiO_2\left(Solid\right) \tag{1}$$

$$Si\left(Solid\right)+2H_2O \rightarrow SiO_2\left(Solid\right)+2H_2 \tag{2}$$

The substrate Si atoms are consumed by the oxidation reaction. The relationship between SiO$_2$ layer thickness d_{ox} and consumed Si substrate layer thickness d_{Si} is derived from the conservation of number of Si atoms of

$$d_{ox}n_{ox} = d_{Si}n_{Si} \tag{3}$$

which gives

$$d_{Si} = \frac{n_{ox}}{n_{Si}}d_{ox} = \alpha d_{ox} = \frac{2.3\times10^{22}\,cm^{-3}}{5.0\times10^{22}\,cm^{-3}}d_{ox} = 0.46d_{ox} \tag{4}$$

where $\alpha = {}^{n_{ox}}\!/\!_{n_{Si}}$.

Furthermore, the oxidation process induces redistribution of impurities in silicon substrates, which influences device characteristics.

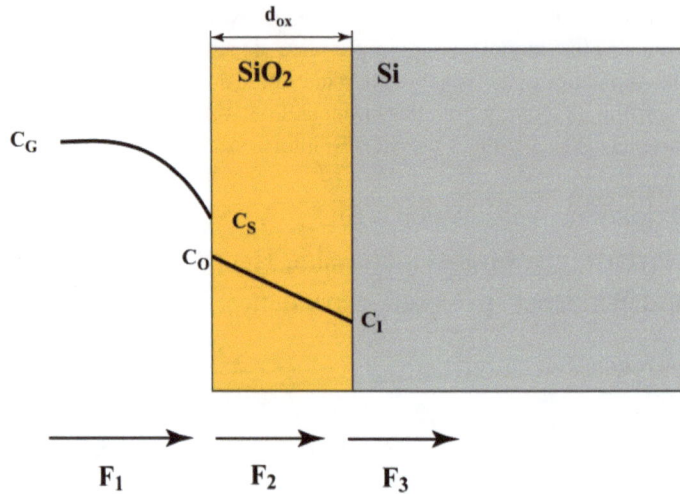

Figure 1: Oxidant flux from the gas phase to the SiO$_2$/Si interface during thermal oxidation.

KINETICS OF OXIDE GROWTH

The kinetics of the oxidation process is studied here. Fig. **1** shows a schematic model that was described by Deal and Grove [1]. The model predicts linear parabolic thickness dependence on oxidation time.

We consider O$_2$ or H$_2$O gas flow to the wafer surface. We assume that the gas concentration far from the wafer surface is constant and is C_G , and the concentration at the wafer surface is C_s . The corresponding flux F_1 in the gas phase is expressed by

$$F_1 = h_G \left(C_G - C_S \right) \tag{5}$$

where h_G is the mass-transfer constant. We assume Henry's law, where the concentration of the solid substrates at the surface is proportional to the partial pressure of the species in the surrounding gas. Therefore, C_G and C_S are expressed as follows.

$$C^* = HP_G \tag{6}$$

$$C_0 = HP_S \tag{7}$$

where H is the Henry's law constant, P_G is the partial pressure of oxidant gas far from the wafer surface, and P_S is the partial pressure of the oxidant gas right to the wafer surface. C_0 is the concentration of oxidant species at the surface of the SiO$_2$ layer, and C^* is the concentration of the oxidant species in the solid when the surface concentration in the gas phase is C_G.

We assume the ideal gas law as

$$C_G = \frac{P_G}{k_B T} \tag{8}$$

$$C_S = \frac{P_S}{k_B T} \tag{9}$$

The concentrations in the gas phase are related to the corresponding concentrations in the solid phase as

$$C_G = \frac{C^*}{H k_B T} \tag{10}$$

$$C_S = \frac{C_0}{H k_B T} \tag{11}$$

Therefore, the flux F_1 is expressed by the concentrations in the solid phase as

$$
\begin{aligned}
F_1 &= \frac{h_G}{H k_B T}\left(C^* - C_0\right) \\
&= h\left(C^* - C_0\right)
\end{aligned}
\tag{12}
$$

where

$$h \equiv \frac{h_G}{H k_B T} \tag{13}$$

Oxidant species diffuse in the existing SiO₂ layer, reach the SiO₂/silicon interface, react with substrate Si atoms, and form SiO₂ layer. We assume that the process is in steady state.

The flux in the oxide layer F_2 is given by

$$F_2 = D\frac{C_s - C_I}{d_{ox}} \tag{14}$$

where D is the diffusivity of the oxidizing species in the oxide layer, and C_I is the concentration at the SiO_2 / Si interface. The reaction at the surface is given by

$$F_3 = k_s C_I \tag{15}$$

where k_s is the reaction constant.

In the steady state, we can assume each flux is balanced and is given by

$$F_1 = F_2 = F_3 \equiv F \tag{16}$$

This gives us the flux at steady state F as

$$F = \frac{k_S C^*}{1 + \dfrac{k_S}{h} + \dfrac{k_S d_{ox}}{D}} \tag{17}$$

The differential equation for forming oxide layer is then given by

$$n_{ox}\frac{d(d_{ox})}{dt} = \frac{k_S C^*}{1 + \dfrac{k_S}{h} + \dfrac{k_S d_{ox}}{D}} \tag{18}$$

We solve the differential equation with the initial condition of

$$d_{ox}(0) = d_i \tag{19}$$

$$\int_{d_i}^{d_{ox}}\left(1 + \frac{k_S}{h} + \frac{k_S d_{ox}}{D}\right)d(d_{ox}) = \frac{k_S C^*}{n_{ox}}\int_o^t dt \tag{20}$$

The integration is performed as

$$\left(1+\frac{k_S}{h}\right)(d_{ox}-d_i)+\frac{1}{2}\frac{k_S}{D}\left(d_{ox}^{\;2}-d_i^{\;2}\right)=\frac{k_S C^*}{n_{ox}}t \tag{21}$$

This can be reduced to

$$d_{ox}^{\;2}+Ad_{ox}-B(t+\tau)=0 \tag{22}$$

where

$$A=\frac{2D}{k_S}\left(1+\frac{k_S}{h}\right) \tag{23}$$

$$B=\frac{2DC^*}{n_{ox}} \tag{24}$$

$$\tau=\frac{d_i^{\;2}+Ad_i}{B} \tag{25}$$

In the derivation process, we introduce a parameter d_i, which expresses the initial oxide thickness before the oxidation. It also empirically expresses anomalous initial oxidation that deviate from the Deal Grove model.

Equation 22 can be solved as

$$d_{ox}=\frac{A}{2}\left|\sqrt{1+\frac{4B(t+\tau)}{A^2}}-1\right| \tag{26}$$

This can be simplified for two limiting cases as

$$d_{ox}=\begin{cases}\dfrac{B}{A}(t+\tau) & for\; t\ll \dfrac{A^2}{2B}\\[2mm]\sqrt{Bt} & for\; t\gg \dfrac{A^2}{2B}\end{cases} \tag{27}$$

Therefore, the oxidation rate changes from linear dependence to parabolic dependence as the oxidation time goes on. The linear rate constant is given by

$$\frac{B}{A} = \frac{\dfrac{2DC^*}{n_{ox}}}{\dfrac{2D}{k_S}\left(1+\dfrac{k_S}{h}\right)}$$

$$= \frac{1}{\left(\dfrac{1}{k_S}+\dfrac{1}{h}\right)}\frac{C^*}{n_{ox}}$$

(28)

The critical time of changing from linear dependence to parabolic dependence t_c is given by

$$t_c = \frac{A^2}{2B}$$

$$= \frac{n_{ox}}{C^*}\left(\frac{1}{k_S}+\frac{1}{h}\right)D$$

(29)

B is proportional to the partial pressure of the oxidant gas. The principal effect of temperature on B is reflected in the diffusion coefficient of oxidant species in SiO_2.

Both B and B/A are well expressed by Arrhenius expressions of the form

$$B = B_0 \exp\left(-\frac{\Delta E_B}{k_B T}\right)$$

(30)

$$\frac{B}{A} = \left(\frac{B}{A}\right)_0 \exp\left(-\frac{\Delta E_{B/A}}{k_B T}\right)$$

(31)

and the related parameter values are listed in Table **1** [2].

Table 1: Parameter values of B and B/A in (100) oriented substrates [2]

Parameter	Ambient	Prefactor $\mu m^2 / hr$	Activation energy eV
B	O_2	7.72×10^2	1.23
	Wet O_2	2.14×10^2	0.71
B/A	O_2	3.71×10^6	2.00
	Wet O_2	5.33×10^7	2.05

The activation energy of parabolic rate B is quite different for dry O_2 and wet O_2 oxidation. This reflects that the diffusion species in SiO_2 are not O but O_2 and H_2O. This means that the molecules are not dissolved in SiO_2.

The activation energy of B/A is almost the same of around 2 eV. The reaction process is not simple. It includes the reactions of oxidant species dissolution resulting O generation, the breaking of Si bonds resulting in Si generation, and the formation of SiO_2. The activation energy of around 2.0 eV corresponds to Si-Si bond breaking process. Therefore, this process is the dominant process in the reaction at the Si/SiO_2 interface. This also suggests that $k_s \ll h$ in Eq. 29, and it reduces to

$$\frac{B}{A} = \frac{k_S C^*}{n_{ox}}$$

(32)

Fig. **2** shows the dependence of oxide thickness on time which is evaluated using Eq. 26. The initial anomalous oxidation process is ignored in the calculation, that is, it is assumed that $d_i = 0$. The dependence changes from linear to the parabolic. The oxidation rate of wet O_2 is much larger than that of dry O_2.

The linear parabolic model described above is widely accepted. However, experimental oxide thickness sometimes deviates from the model, in particular for very thin oxides (<20nm). Many models have been proposed to explain the deviation, but none of them has a wide spread acceptance [2].

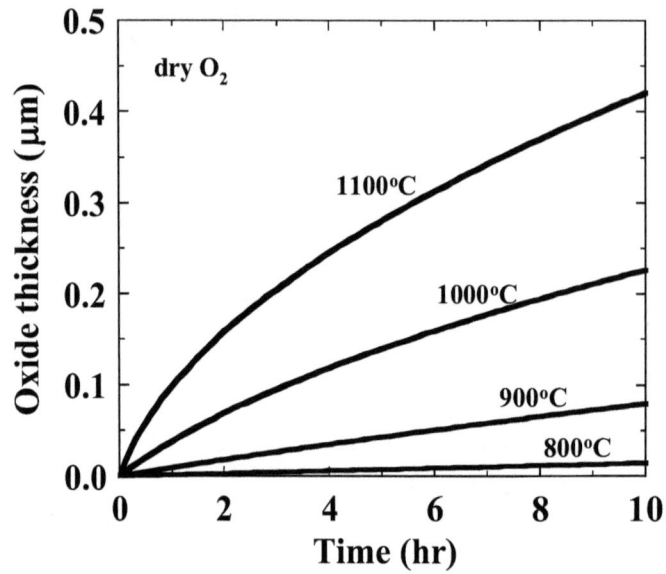

(a)

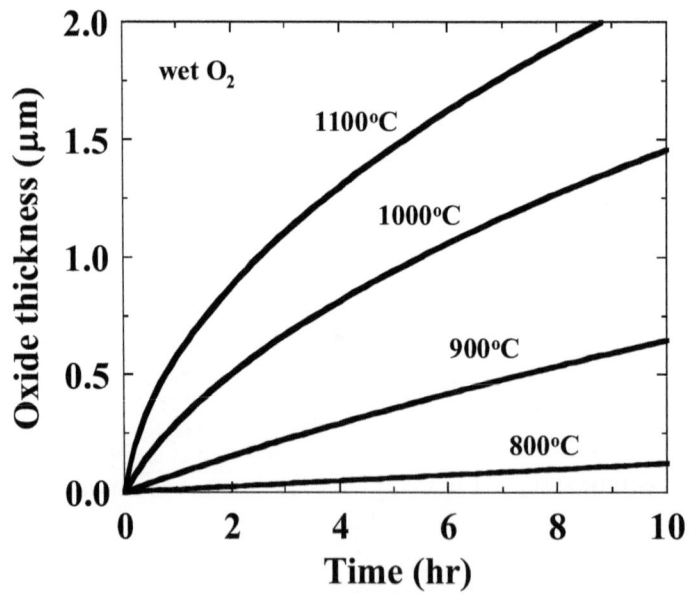

(b)

Figure 2: Dependence of oxide thickness on time for various temperatures. (a) Dry oxidation, (b) Wet oxidation.

Among them, the Massoud model [3] is used in all commercial simulator, and is given by

$$\frac{d\left(d_{ox}\right)}{dt} = \frac{B}{2d_{ox}+A} + C\exp\left(-\frac{d_{ox}}{L}\right) \tag{33}$$

where

$$C = C_0 \exp\left(-\frac{\Delta E_L}{k_B T}\right) \tag{34}$$

$C_0 = 3.6 \times 10^8 \; ^{\mu m}/_{hr}$, $\Delta E_L = 2.35 \, eV$, and $L = 7 \, nm$. The first term in Eq. 33 is the Deal Grove model, and the second term represents the additional oxidation mechanism. The term plays a role to increase d_{ox} in the early stage of oxidation process around $d_{ox} = L$,

One pragmatic model is to express the dependence with a simple power law as

$$d_{ox} = at^b \tag{35}$$

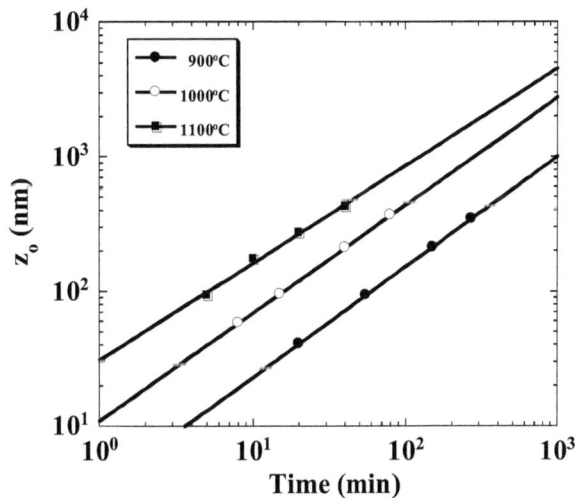

Figure 3: Dependence of experimental oxide thickness on dry oxidation time.

Fig. **3** summarizes the dependence of the oxide thickness z0 on time, where the substrates were wet-oxidized at $900^{\circ}C$, $1000^{\circ}C$, and $1100^{\circ}C$, which is empirically expressed by [4]

$$z_0\left[cm\right] = \begin{cases} 1.22\times10^{-8}t^{0.82} & \textit{for } 900^{\circ}C \\ 4.16\times10^{-8}t^{0.80} & \textit{for } 1000^{\circ}C \\ 1.61\times10^{-7}t^{0.72} & \textit{for } 1100^{\circ}C \end{cases} \tag{36}$$

It should be noted that the dependences of z_0 on time deviate from the parabolic one in the practical experimental conditions.

REDISTRIBUTION OF IMPURITIES DURING THERMAL OXIDATION

The B concentration at the Si-SiO$_2$ interface on the Si side influences the threshold voltage and leakage current. Therefore, predicting the concentration after a certain oxidation is very important. The redistribution model has been derived by Grove *et al.* assuming parabolic oxidation dependence [5], and the model is extended to more general oxidation dependence [4].

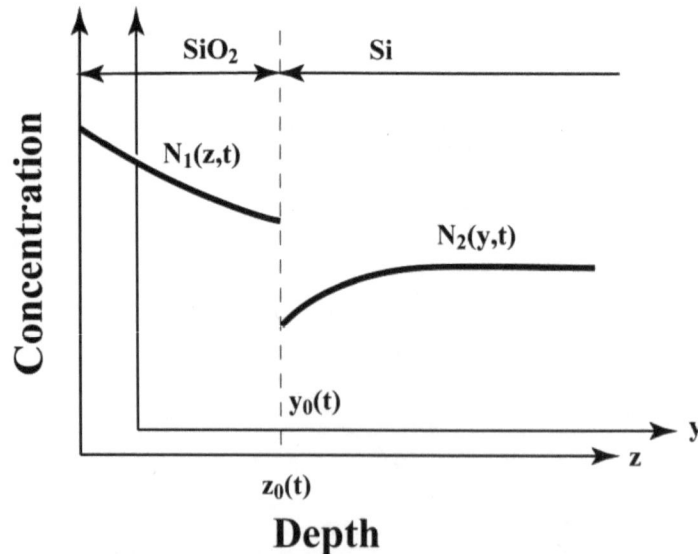

Figure 4: Axis for the analytical model. The depth in the SiO$_2$ layer is expressed by z, where the origin is at the SiO$_2$ surface. The thickness of the SiO$_2$ layer is given by z$_0$. The depth in the Si layer is expressed by y, where the origin is at the initial silicon surface, and y$_0$ corresponds to the Si layer thickness that is the consumed due to the oxidation.

We define two axis systems regarding to Si layer and growing SiO_2 layer as shown in Fig. **4**. z denotes the depth in the SiO_2 layer, and z_0 corresponds the oxide thickness at the time. y denotes the depth in Si layer, and y_0 corresponds to the Si layer thickness consumed by the oxidation.

We must solve the diffusion equation in SiO_2 and Si given by

$$\frac{\partial N_1}{\partial t} = D_1 \frac{\partial^2 N_1}{\partial z^2} \ for \ 0 < z < z_0 \tag{37}$$

$$\frac{\partial N_2}{\partial t} = D_2 \frac{\partial^2 N_2}{\partial z^2} \ for \ y_0 < y < \infty \tag{38}$$

where N_1 is the B concentration in SiO_2 layer and D_1 is the corresponding diffusion coefficient, and N_2 is the B concentration in Si layer and D_2 is the corresponding diffusion coefficient.

We treat the initial uniform doping concentration in Si layer, and the boundary conditions that impose continuity of flux and thermal equilibrium of segregation at the SiO_2/Si interface, which are given by

$$N_2(y,0) = N_B \tag{39}$$

$$-D_1 \frac{\partial N_1}{\partial z}\bigg|_{z=0} = -hN_1(0,t) \tag{40}$$

$$N_2(\infty,t) = N_B \tag{41}$$

$$N_2(y_0,t) = mN_1(z_0,t) \tag{42}$$

h is the surface transport coefficient, and m is the segregation coefficient.

The diffusion coefficient in SiO_2 is quite small as shown in Fig. **5**, and we neglect the flux in SiO_2 layer. We then simplify the diffusion equation 37 as

$$\frac{\partial N_1}{\partial t} = 0 \tag{43}$$

This also means that we can neglect out diffusion, that is, we do not need to consider the boundary condition of Eq. 40.

We need a boundary condition for flux at the SiO$_2$/Si interface. However, the interface moves during oxidation, and we cannot express the continuity of flux in this case, which are investigated as the following [4, 5]. The total net boron in SiO$_2$ and Si layers, Q, is

$$Q = \int_0^{z_0} N_1(z)dz + \int_{y_0}^{\infty} N_2(y,t)dy \tag{44}$$

Q is time invariant since we can neglect out diffusion, which implies

$$\frac{dQ}{dt} = \frac{d}{dt}\int_0^{z_0} N_1(z)dz + \frac{d}{dt}\int_{y_0}^{\infty} N_2(y,t)dy = 0 \tag{45}$$

We introduce the integral functions F_1 and F_2, which are defined by

$$\int_0^{z_0} N_1(z,t)dz = F_1(z_0) - F_1(0) \tag{46}$$

$$\int_{y_0}^{\infty} N_2(y,t)dz = F_2(\infty) - F_1(y_0) \tag{47}$$

The time derivative of Eqs. 46 and 47 are then given by

$$\begin{aligned}
\frac{d}{dt}\int_0^{z_0} N_1(z,t)dz &= \frac{d}{dt}\left[F_1(z_0) - F_1(0)\right] \\
&= \frac{\partial F_1(z_0,t)}{\partial z_0}\frac{dz_0}{dt} \\
&= N_1(z_0,t)\frac{dz_0}{dt}
\end{aligned} \tag{48}$$

$$\begin{aligned}
\frac{d}{dt}\int_{y_0}^{\infty} N_2(y,t)dz &= \frac{d}{dt}\left[F_2(\infty) - F_1(y_0)\right] \\
&= \frac{\partial F_2(\infty,t)}{\partial t} - \frac{\partial F_1(y_0,t)}{\partial t} - \frac{\partial F_1(y_0,t)}{\partial y_0}\frac{dy_0}{dt} \\
&= \int_{y_0}^{\infty} \frac{\partial N_2(y,t)}{\partial t}dy - N_2(y_0,t)\frac{dy_0}{dt} \\
&= \int_{y_0}^{\infty} D_2\frac{\partial^2 N_2(y,t)}{\partial y^2}dy - N_2(y_0,t)\frac{dy_0}{dt} \\
&= -D_2\frac{\partial N_2(y,t)}{\partial y}\bigg|_{y=y_0} - N_2(y_0,t)\frac{dy_0}{dt}
\end{aligned} \tag{49}$$

Substituting Eqs. 49 and 50 into 46, we obtain

$$N_1\left(z_0,t\right)\frac{dz_0}{dt} - D_2\left.\frac{\partial N_2\left(y,t\right)}{\partial y}\right|_{y=y_0} - N_2\left(y_0,t\right)\frac{dy_0}{dt} = 0 \tag{50}$$

The consumed Si thickness is related to the growing SiO$_2$ thickness as

$$y_0\left(t\right) = \alpha z_0\left(t\right) \tag{51}$$

Substituting Eqs. 52 and 42 into Eq. 51, we obtain a boundary condition associated with SiO$_2$/Si interface as

$$\left(\frac{1}{m} - \alpha\right)\frac{dz_0}{dt}N_2\left(y_0,t\right) = D_2\left.\frac{\partial N_2\left(y,t\right)}{\partial y}\right|_{y=y_0} \tag{52}$$

This boundary condition can be derived with a different way considering physical meaning as follows.

Si layer is expanded by oxidation, and hence the interface impurity concentration in Si $N_2\left(y_0,t\right)$ is decreased by the expansion, and is given by

$$N_2\left(y_0,t\right)\frac{dy_0}{dz_0} = \alpha N_2\left(y_0,t\right) \tag{53}$$

We implicitly assumed that there is no impurity flux in this growing oxide region. However, flux should exist to establish the thermal equilibrium for the segregation, the concentration in the growing oxide layer must be

$$\frac{1}{m}N_2\left(y_0,t\right) \tag{54}$$

This means that a flow during the time period dt is given by

$$\left[\frac{1}{m}N_2\left(y_0,t\right) - \alpha N_2\left(y_0,t\right)\right]dz_0 \tag{55}$$

Therefore, the continuity of the flux at the interface is given by the following

$$\left[\frac{1}{m}N_2\left(y_0,t\right)-\alpha N_2\left(y_0,t\right)\right]dz_0 = D\frac{\partial N_2\left(y,t\right)}{\partial y}\bigg|_{y=y_0} dt \tag{56}$$

This is exactly the same as Eq. 53.

We introduce a variable of

$$\eta = \frac{y}{2\sqrt{D_2 t}} \tag{57}$$

The corresponding concentration variable v_2 is given by

$$N_2\left(z,t\right) = v_2\left(\eta,t\right) \tag{58}$$

The differential Eq. 38 becomes

$$4t\frac{\partial v_2\left(\eta,t\right)}{\partial t} = \frac{\partial^2 v_2\left(\eta,t\right)}{\partial \eta^2} + 2\eta\frac{\partial v_2\left(\eta,t\right)}{\partial \eta} \tag{59}$$

We impose the form of the solution of v_2 as

$$v_2\left(\eta,t\right) = T\left(t\right)G\left(\eta\right) \tag{60}$$

Substituting Eq. 61 into 60, we obtain

$$4t\frac{T'}{T} = \frac{G''}{G} + 2\eta\frac{G'}{G} \tag{61}$$

The variables of t and η are independent, and hence both sides of Eq. 61 should not depend on the equal of the variables. We set the constant number to be λ, and reduce Eq. 61 to

$$\begin{cases} \dfrac{T'}{T} = \dfrac{\lambda}{4t} \\ G'' + 2\eta G' - \lambda G = 0 \end{cases} \tag{62}$$

The first part of Eq. 62 gives a time dependent solution of

$$T = T_0 t^{\frac{\lambda}{4}} \tag{63}$$

Where T_0 is an arbitrary constant. Note that v_2 corresponds to the concentration over the region of Si layer. Positive λ leads to infinite concentration, and negative λ leads to zero concentration for infinite time, which are inadequate. Therefore, λ must be zero, and second part of Eq. 62 is reduced to

$$G'' + 2\eta G' = 0 \tag{64}$$

which is solved as

$$v_2(\eta, t) = C_1 erf\eta + C_2 \tag{65}$$

where C_1 and C_2 are arbitrary constants. Then, we obtain

$$N_2(y,t) = C_1 erf\left(\frac{t}{2\sqrt{D_2 t}}\right) + C_2 \tag{66}$$

Using the boundary condition of Eq. 41, Eq. 66 becomes

$$N_2(y,t) = N_B - C_1 erfc\left(\frac{y}{2\sqrt{D_2 t}}\right) \tag{67}$$

Then, the interface concentration is given as

$$N_2(y_0) = N_B - C_1 erfc\left(\frac{y_0}{2\sqrt{D_2 t}}\right) \tag{68}$$

Eliminating C_1 from Eqs. 67 and 68, we obtain

$$\frac{N_2(y,t) - N_B}{N_2(y_0,t) - N_B} = \frac{erfc\left(\frac{y}{2\sqrt{D_2 t}}\right)}{erfc\left(\frac{y_0}{2\sqrt{D_2 t}}\right)} \tag{69}$$

Substituting Eq. 69 into Eq. 53, $N_2(y_0, t)$ is given by

$$N_2(y_0, t) = \frac{N_B}{1 + \sqrt{\frac{\pi t}{D_2}}\left(\frac{1}{m} - \alpha\right) erfc\left(\frac{y_0}{2\sqrt{D_2 t}}\right) \exp\left(\frac{y_0^2}{4D_2 t}\right) \frac{dz_0}{dt}} \tag{70}$$

Since we can neglect the diffusion in the SiO_2 layer, we can obtain the impurity concentration at z in SiO_2 from the interface concentration when z_0 is z dividing by m, which is given by

$$N_1(z, t) = \frac{N_B}{m} \frac{1}{1 + \sqrt{\frac{t(z)}{D_2}}\left(\frac{1}{m} - \alpha\right) erfc\left(\frac{\alpha z}{2\sqrt{D_2 t(z)}}\right) \exp\left(\frac{\alpha^2 z^2}{4D_2 t(z)}\right) \frac{dz_0}{dt}\Big|_{z_0 = z}} \tag{71}$$

The time dependence of oxide thickness is empirically expressed by

$$z_0 = At^r \tag{72}$$

The arbitrary constants A and r are determined by fitting the experimental data. Therefore, $t(z)$, and $\frac{dz_0}{dt}$ for $z_0 = z$ are given by

$$t(z) = \left(\frac{z}{A}\right)^{\frac{1}{r}} \tag{73}$$

$$\frac{dz_0}{dt}\Big|_{z_0 = z} = rAt(z)^{r-1}$$

$$= rA\left(\frac{z}{A}\right)^{\frac{r-1}{r}} \tag{74}$$

Fig. **5** compares the experimental and theoretical results. The analytical model well reproduces the experimental data for various temperatures.

As the oxidation continues, the interface concentration decreases as shown in Fig. **6**. This is a consequence of the fact that the dependence of the oxide thickness on time is greater than a power of 0.5. The silicon surface concentration simply decreases with proceeding oxidation.

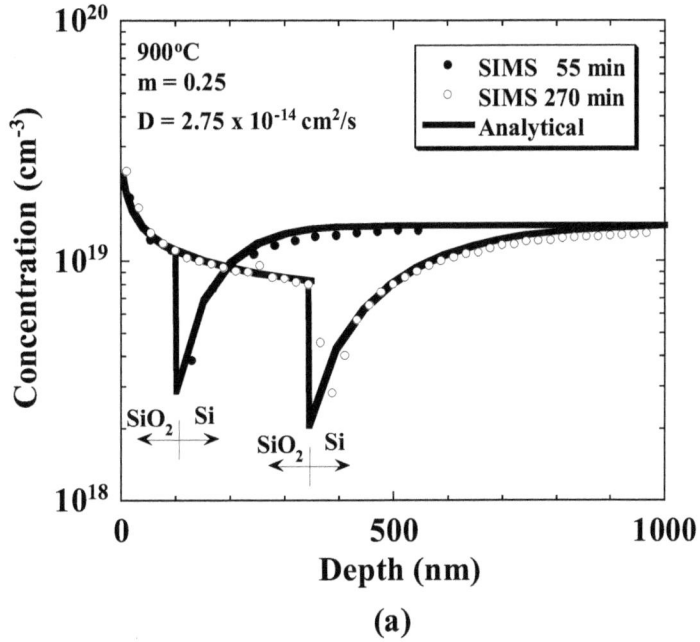

(a)

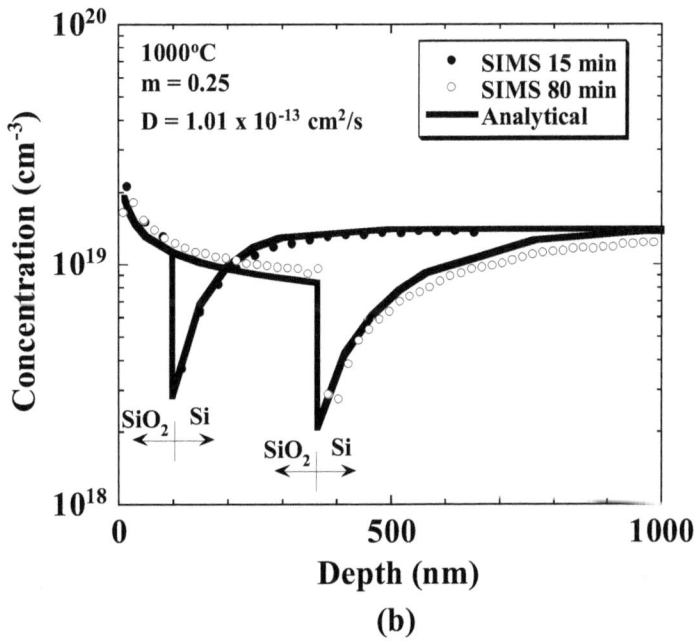

(b)

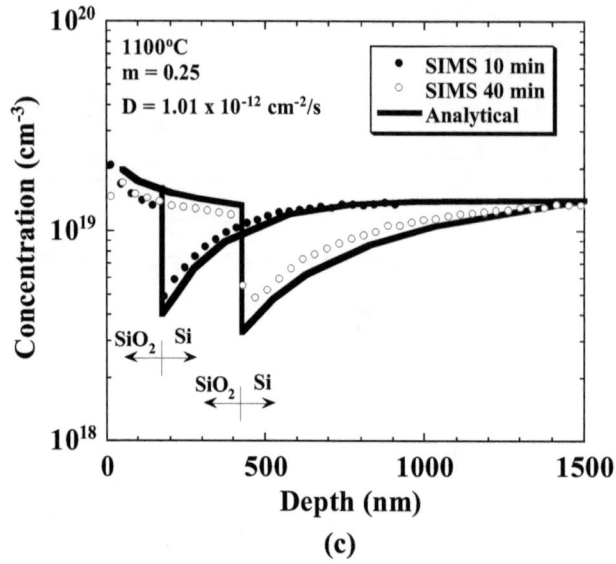

(c)

Figure 5: Experimental and analytical redistribution boron concentration profile in SiO_2 and Si during oxidation. (a) 900°C, (b) 1000°C, and (c) 1100°C.

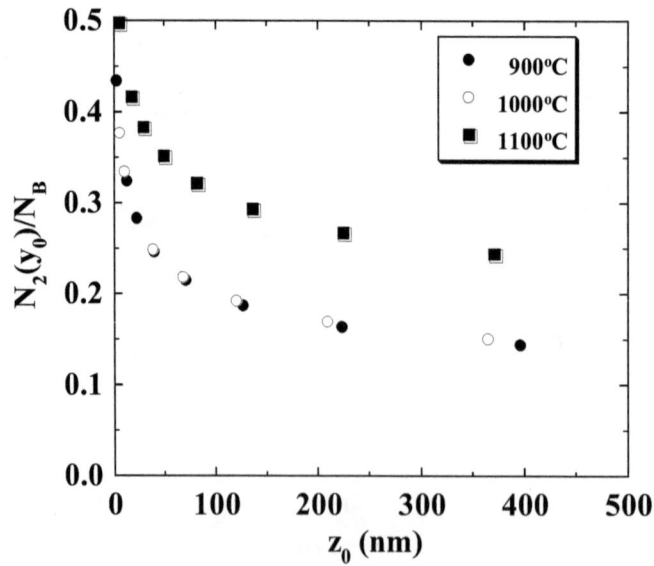

Figure 6: Dependence of the concentration at the silicon surface on oxidation thickness.

REFERENCES

[1] A. S. Grove, Physics and Technology of Semiconductor Devices, John Wiley & Sons, New York, 1967, p. 22.

[2] J. D. Plummer, M. D. Deal, and P. B. Griffin, Silicon VLSI Technology, Prentice Hall, New Jersey, 2000, p. 287.

[3] H. Z. Massoud, J. D. Plummer, and E. A. Irene, "Thermal oxidation of silicon in dry oxygen: Growth-rate enhancement in the thin regime I. Experimental results, II. Physical mechanisms," J. Electrochem. Soc., vol. 132, pp. 2685-2693, 1985.

[4] K. Suzuki and T. Miyashita, "Models of boron redistribution during thermal oxidation with general oxidation rate," IEEE Trans. Electron Devices, ED-47, pp. 523-528, 2000.

[5] A. S. Grove, O. Leistiko, and C. T. Sah, "Redistribution of acceptor and donor impurities during thermal oxidation of silicon," J. Appl. Phys., vol. 35, pp. 2695-2701, 1964.

CHAPTER 6

Segregation

Abstract: Segregation is a coefficient defined as the ratio of impurity concentrations at both sides of two different layers, which influences the diffusion profiles. However, the ratio is rarely in thermal equilibrium in general cases. Hence, the evaluation of the value of the segregation is difficult. It is found that the thermal equilibrium of segregation has been established in the redistribution profile of impurities in oxidized polycrystalline Si (polysilicon) because the diffusion coefficient is much larger than that in Si. The redistribution model is derived, and related segregation values are evaluated.

Keywords: Segregation, thermal equilibrium, polycrystalline silicon, diffusion, redistribution, chemical potential, interface, Si/SiO$_2$ interface, oxidation, transport coefficient, chemical vapor deposition, B, As, SIMS, SiO$_2$/polysilicon interface, absorption coefficient, dose.

INTRODUCTION

Segregation is related to the chemical potential of impurities in substrate materials, and it is defined as the ratio of impurity concentrations at both sides of two different layers. The outstanding important system is the interface between Si and SiO$_2$, where the impurity concentration in Si near the Si/SiO$_2$ interface is influenced by the diffusion coefficients in both layer and the segregation coefficient. Hence, the impurity profiles near the Si/SiO$_2$ interface in oxidation process draws much attention. However, the redistribution profiles are affected by many factors, such as segregation coefficient, segregation transport coefficient, diffusion in Si and in SiO$_2$ and oxidation rate. Furthermore, the experimental evaluation of the impurity concentration at both sides of the interface is difficult.

It is found that the thermal equilibrium of segregation has been established in the redistribution profile of impurities in oxidized polycrystalline Si (polysilicon). The reason for this is that the diffusion coefficient is much larger than that in Si. The redistribution model is derived, and related values are evaluated [1, 2].

Kunihiro Suzuki

EXPERIMENTAL PROFILES DURING POLYSILICON OXIDATION

An 800-nm-thick polysilicon layer was deposited on Si_3N_4 layer by chemical vapor deposition, and B was ion implanted at 60 keV with a dose of 2 x 10^{16} cm^{-2}, or As at 150 keV with a dose of 1 x 10^{16} cm^{-2}. After annealing at 900°C for 30 min to flatten the impurity profiles in polysilicon, samples were wet-oxidized at various temperatures.

Fig. **1** shows SIMS B profiles after oxidation at 1000°C for 70 and 140 min. The B concentration decreases gradually with depth in SiO_2. A discontinuity associated with segregation occurs in the B concentration at the SiO_2-polysilicon interface. The B profile in polysilicon is uniform due to large diffusion coefficient in the layer. The B profile in the 70-min oxidized SiO_2 is exactly the same as that in 140-min oxidized SiO_2. Thus, B redistribution in SiO_2 during oxidation is negligible.

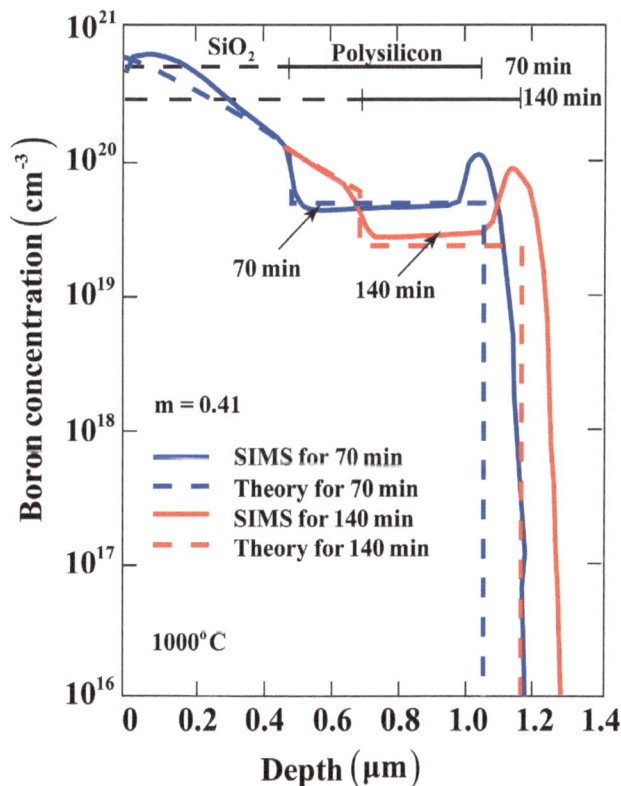

Figure 1: SIMS B concentration profiles. Theoretical profiles are also shown.

Fig. **2** shows the ration of N_2/N_1 at the SiO$_2$ / polysilicon interface, where N_1 is the impurity concentration in SiO$_2$ and N_2 that in polysilicon. The ratio does not vary during oxidation, and the thermal equilibrium in segregation is established.

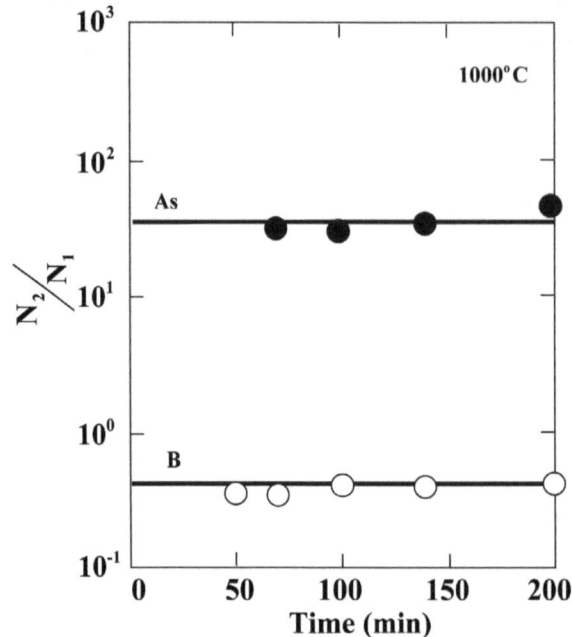

Figure 2: Ratio of impurity concentration in polysilicon (N_2) and to that in SiO$_2$ (N_1) at the SiO$_2$-polysilicon interface.

Fig. **3** shows the SIMS data of As concentration profiles after oxidation for 70 and 140 min. The As concentration profiles differ from the B profiles. However, the general trends are similar: negligible redistribution in SiO$_2$, uniform impurity concentration in polysilicon, and invariable N_2/N_1.

As oxidation proceeds, polysilicon is consumed and polysilicon thickness decreases. Fig. **4** shows the dependence of impurity concentration in polysilicon on decreased polysilicon thickness due to oxidation. The B concentration decreases drastically with decreasing polysilicon thickness because B in the residual polysilicon region is absorbed in the growing SiO$_2$ region. In contract, the As concentration in polysilicon increases slightly with decreased polysilicon thickness because As in the growing SiO$_2$ erupts into the residual polysilicon region.

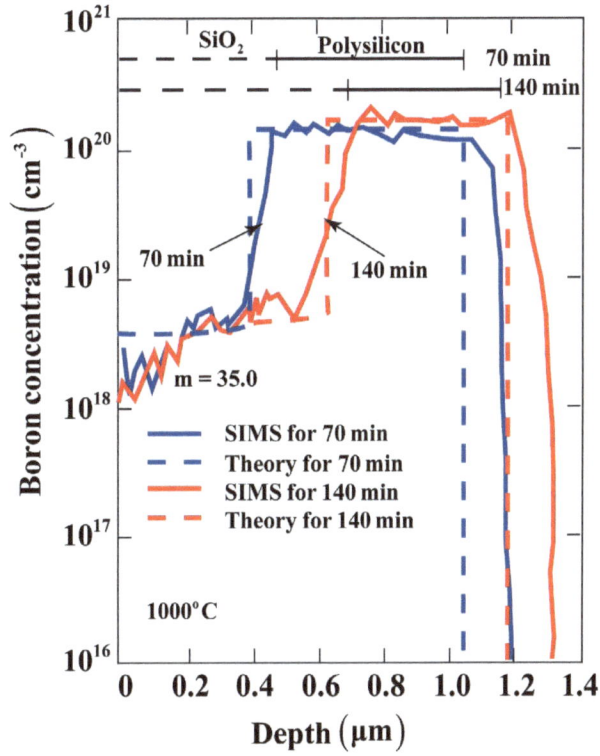

Figure 3: SIMS As concentration profiles. Theoretical profiles are also shown.

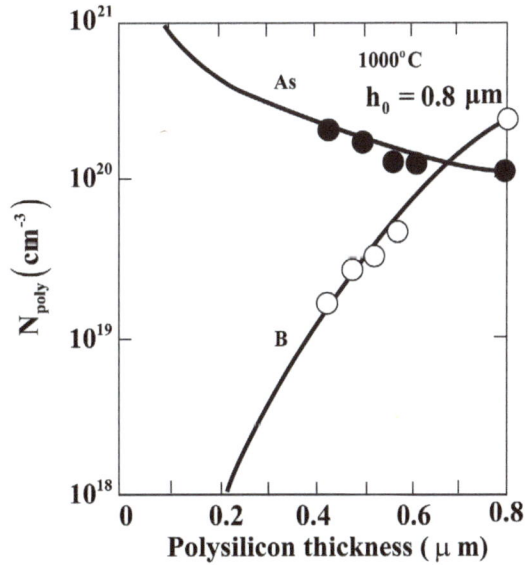

Figure 4: Dependence of impurity concentration on residual polysilicon thickness. Theoretical data are also shown.

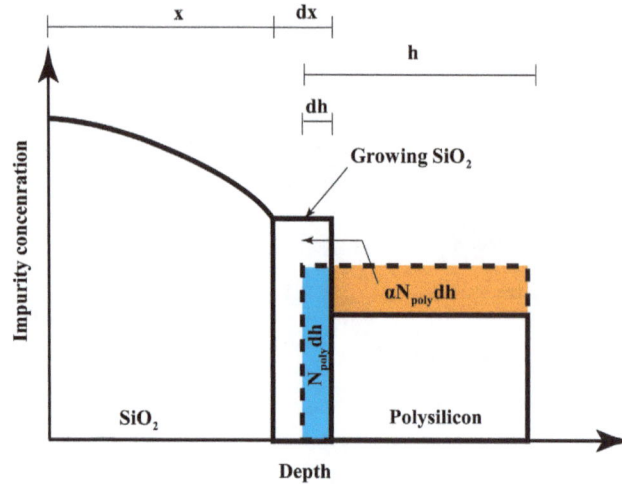

Figure 5: Model for forming the redistribution profile. h-thick polysilicon is oxidized. Polysilicon dh thick is consumed and dx thick SiO$_2$ is grown. The total net impurity in the grown SiO$_2$ is those existed in the thick polysilicon $N_{poly}dh$ and those absorbed from the residual polysilicon $\alpha N_{poly}dh$, and the thermal equilibrium of segregation is established in the growing SiO$_2$ and residual polysilicon system.

THEORY

Inspecting the above experimental data, we assume the followings to derive a model for the redistribution of the impurities.

(i) The diffusion coefficients in SiO$_2$ is zero because the impurity concentration profiles in SiO$_2$ is invariant during oxidation.

(ii) Diffusion coefficient in polysilicon is infinite because the profiles in polysilicon are always flat.

(iii) The thermal equilibrium of segregation is established because the ratio of N_1/N_2 is unchanged during oxidation.

Fig. **5** shows the explanation of the model.

As oxidation proceeds, SiO$_2$ thickness increases from x to $x+dx$ and polysilicon thickness decreases from h to $h-dh$. The net decreased impurities in polysilicon $-dQ$ is expressed by

$$-dQ = N_{poly} v dt + \alpha N_{poly} v dt \tag{1}$$

where N_{poly} is the impurity concentration in polysilicon, and v is the decreasing speed of polysilicon and is given by

$$v = -\frac{dh}{dt} \tag{2}$$

The first component of Eq. 1 is net impurities existed in the consumed polysilicon region, and the second component is net impurities incorporated from the residual polysilicon region to establish the thermal equilibrium of segregation. α is an absorption coefficient later related to the segregation coefficient.

Substituting Eq. 2 into Eq. 1, we obtain

$$-dQ = -\left(N_{poly} dh + \alpha N_{poly} dh \right) \tag{3}$$

$-dQ$ is also expressed in general as

$$\begin{aligned} -dQ &= -\left(N_{poly} h \right) \\ &= -\left(h dN_{poly} + N_{poly} dh \right) \end{aligned} \tag{4}$$

Combining Eqs. 3 and 4, we obtain

$$N_{poly} = N_0 \left(\frac{h}{h_0} \right)^{\alpha} \tag{5}$$

where N_0 is the initial impurity concentration in polysilicon defined as

$$N_0 = \frac{\Phi}{h_0} \tag{6}$$

h_0 is the initial polysilicon thickness, and Φ is the implanted dose of impurities.

Net decreased impurities in polysilicon for Eq. 3 are the net impurities incorporated into the growing SiO_2 region. The thickness increases when polysilicon is changed to SiO_2 in oxidation by a factor of r (=2.17), that is

$$dx = rdh \tag{7}$$

as shown in Fig. **5**. Substituting Eq. 7 into Eq. 3, we obtain

$$dQ = \frac{1+\alpha}{r} N_{poly} dx \tag{8}$$

dQ is also expressed in general as

$$dQ = N_{SiO_2} dx \tag{9}$$

where N_{SiO_2} is the impurity concentration in the growing SiO_2 at polysilicon thickness h. Combining Eq. 8 and 9, N_{SiO_2} becomes

$$N_{SiO_2} = \frac{1+\alpha}{r} N_0 \left(\frac{h}{h_0}\right)^{\alpha} \tag{10}$$

x and h are related to each other as

$$r(h_0 - h) = x \tag{11}$$

Substituting Eq. 11 into Eq. 10, N_{SiO_2} at x is solved as

$$N_{SiO_2} = \frac{1+\alpha}{r} N_0 \left(1 - \frac{x}{rh_0}\right)^{\alpha} \tag{12}$$

The thermal equilibrium of segregation is established, then

$$m = \frac{N_{poly}}{N_{SiO_2}\left(r(h_0 - h)\right)} \tag{13}$$

Thus, substituting Eqs. 5 and 12 into Eq. 13, α is related to m as

$$\alpha = \frac{r}{m} - 1 \tag{14}$$

The impurity profile in both SiO_2 and polysilicon is then

$$N(x) = \begin{cases} \dfrac{1+\alpha}{r} N_0 \left(1 - \dfrac{x}{rh_0}\right)^{\alpha} & 0 \leq x \leq r(h_0 - h) : in\ SiO_2 \\ N_0 \left(\dfrac{h}{h_0}\right)^{\alpha} & r(h_0 - h) \leq x \leq r(h_0 - h) + h : in\ polysilicon \end{cases} \quad (15)$$

The theory well reproduces the experimental data as shown in Figs. **1**, **3**, and **4**. According to the theory, m does not determine directly whether impurities in the residual polysilicon region are absorbed in the growing SiO_2 region or impurities in growing SiO_2 region erupt into the residual polysilicon region. This is determined by the sign of α, which is positive when m is larger than r.

We can simply determine the segregation coefficient just by evaluating the concentration in polysilicon with Eq. 5.

We preformed similar experiments for various temperatures and evaluated the corresponding segregation coefficients. We denote the segregation coefficient of B as m_B and that of As as m_{As}. The data are fitted as

$$m_B = 0.357 \exp\left[-\frac{0[eV]}{k_B T}\right] \quad (16)$$

$$m_{As} = 4.60 \times 10^4 \exp\left[-\frac{0.73[eV]}{k_B T}\right] \quad (17)$$

REFERENCES

[1]   K. Suzuki and Y. Kataoka, "Redistribution of impurities during thermal oxidation of polycrystalline silicon," J. Electrochem. Society, vol. 138, pp. 1794-1798, 1991.

[2]   K. Suzuki, Y. Yamashita, Y. Kataoka, K. Yamazaki, and K. Kawamura, "Segregation coefficient of boron and arsenic at polycrystalline silicon/SiO2 interface," J. Electrochem. Society, vol. 140, pp. 2960-2964, 1993.

Send Orders for Reprints to reprints@benthamscience.net
128 Ion Implantation and Activation, Vol. 3, 2013, 128-204

CHAPTER 7

Analytical Diffusion Profiles Under Various Boundary Conditions

Abstract: We can obtain analytical models for impurity diffusion profiles in some specific cases, although it is difficult to solve this in general. We first show various analytical models assuming constant diffusion coefficients, and then present the models with variable diffusion coefficients. We further display analytical models for two-dimensional diffusion profiles, and profiles for multi-layers.

Keywords: Diffusion coefficient, diffusion length, diffusion equation, differential equation, initial condition, boundary condition, Fourier transformation, Laplace transformation, Gaussian profile, surface, third order moment, forth order moment, Gamma function, Beta function, analytical model.

INTRODUCTION

We hardly obtain analytical diffusion models in general cases, and the profiles are obtained by numerical calculation. However, we can obtain analytical models in some special cases. The analytical models give us clear physical intuition about the diffusion profiles. These models are also utilized to extract physical parameters, for example, diffusion coefficients. Mathematical solutions of the differential equations for diffusion are systematically shown in [1, 2]. We collect examples related to Very Large Scale Integration (VLSI) processes in this chapter.

DIFFUSION PROFILE (CONSTANT DIFFUSION COEFFICIENT)

Constant Amount of Impurities

The diffusion equation is given by

$$\frac{\partial N(x, t)}{\partial t} = D \frac{\partial^2(x, t)}{\partial x^2} \tag{1}$$

where N is the impurity concentration, and D is the diffusion coefficient. The initial condition is given by

$$N(x,0) = f(x) \tag{2}$$

Performing Fourier transformation to the diffusion equation, we obtain

$$\frac{d\mathscr{F}\{N\}}{dt} = D\xi^2 \mathscr{F}\{N\} \tag{3}$$

This can be solved as

$$\mathscr{F}\{N\} = A(\xi)\exp\left(-D\xi^2 t\right) \tag{4}$$

Performing Fourier transformation to the initial condition, we obtain

$$\mathscr{F}\{N(x,0)\} = \mathscr{F}\{f(x)\}\exp\left(-D\xi^2 t\right) \tag{5}$$

Substituting Eq. 5 into Eq. 4, we obtain

$$\begin{aligned}
\mathscr{F}\{N\} &= \mathscr{F}\{f(x)\}\exp\left(-D\xi^2 t\right) \\
&= \mathscr{F}\{f(x)\}\mathscr{F}\left\{\frac{1}{\sqrt{2Dt}}\exp\left(-\frac{x^2}{4Dt}\right)\right\}
\end{aligned} \tag{6}$$

Performing invers Fourier transformation of Eq. 6, we obtain

$$N(x,t) = \frac{1}{\sqrt{2\pi}}\int_{-\infty}^{\infty} f(\xi)\frac{1}{\sqrt{2Dt}}\exp\left(-\frac{(x-\xi)^2}{4Dt}\right)d\xi \tag{7}$$

Let us consider a case where initial profile $f(x)$ of Gaussian given by

$$f(x) = \frac{\Phi}{\sqrt{2\pi}\Delta R_p}\exp\left|-\frac{(x-R_p)^2}{2\Delta R_p^2}\right| \tag{8}$$

Substituting Eq. 8 into Eq. 7, we obtain

$$\begin{aligned}
N(x,t) &= \frac{1}{\sqrt{2\pi}}\int_{-\infty}^{\infty} \frac{\Phi}{\sqrt{2\pi}\Delta R_p}\exp\left[-\frac{(\xi-R_p)^2}{2\Delta R_p^2}\right]\frac{1}{\sqrt{2Dt}}\exp\left(-\frac{(x-\xi)^2}{4Dt}\right)d\xi \\
&= \frac{\Phi}{2\pi\sqrt{2Dt}\Delta R_p}\int_{-\infty}^{\infty} \exp\left[-\left[\frac{(\xi-R_p)^2}{2\Delta R_p^2}+\frac{(x-\xi)^2}{4Dt}\right]\right]d\xi
\end{aligned} \tag{9}$$

The component in the exponential term in Eq. 9 is modified as

$$\frac{\left(\xi-R_p\right)^2}{2\Delta R_p^{\,2}}+\frac{\left(x-\xi\right)^2}{4Dt}=\frac{\left(\left(\xi-x\right)+\left(x-R_p\right)\right)^2}{2\Delta R_p^{\,2}}+\frac{\left(x-\xi\right)^2}{4Dt}$$

$$=\left(\frac{1}{2\Delta R_p^{\,2}}+\frac{1}{4Dt}\right)\left(x-\xi\right)^2-2\frac{x-R_p}{2\Delta R_p^{\,2}}\left(x-\xi\right)+\frac{\left(x-R_p\right)^2}{2\Delta R_p^{\,2}}$$

$$=\frac{\Delta R_p^{\,2}+2Dt}{4\Delta R_p^{\,2}Dt}\left[\left(x-\xi\right)-\frac{2Dt\left(x-R_p\right)}{\Delta R_p^{\,2}+2Dt}\right]^2+\frac{\left(x-R_p\right)^2}{2\left(\Delta R_p^{\,2}+2Dt\right)}$$

We can then obtain

$$N\left(x,t\right)=\frac{\Phi}{2\pi\sqrt{2Dt}\Delta R_p}\sqrt{\frac{\pi 4\Delta R_p^{\,2}Dt}{\Delta R_p^{\,2}+2Dt}}\exp\left|-\frac{\left(x-R_p\right)^2}{2\left(\Delta R_p^{\,2}+2Dt\right)}\right|$$

$$=\frac{\Phi}{\sqrt{2\pi\left(\Delta R_p^{\,2}+2Dt\right)}}\exp\left[-\frac{\left(x-R_p\right)^2}{2\left(\Delta R_p^{\,2}+2Dt\right)}\right]$$

(10)

This is also a Gaussian function which is transformed from initial one to the diffusion one with the moment transformations of

$$\begin{cases}R_p\rightarrow R_p\\[2mm]\Delta R_p\rightarrow\sqrt{\Delta R_p^{\,2}+2Dt}\end{cases}$$

(11)

The profile of Eq. 10 can be derived in a different way. We virtually set the initial condition as a delta function located at $x=R_p$, which is given by

$$f\left(x\right)=\Phi\delta\left(x-R_p\right)$$

(12)

Substituting Eq. 12 into Eq. 7, we obtain

$$N(x, t) = \frac{1}{\sqrt{2\pi}} \int_{-\infty}^{\infty} \Phi \, \delta(\xi - R_p) \frac{1}{\sqrt{2Dt}} \exp\left[-\frac{(x - R_p)^2}{4Dt}\right] d\xi$$

$$= \frac{\Phi}{2\sqrt{\pi Dt}} \exp\left[-\frac{(x - R_p)^2}{4Dt}\right] \tag{13}$$

We regard that the real initial condition is the profile of Eq. 13 after diffusion of a certain time period of t_0. Comparing both of them, we obtain

$$\frac{\Phi}{\sqrt{2\pi \Delta R_p^2}} \exp\left[-\frac{(x - R_p)^2}{2\Delta R_p^2}\right] = \frac{\Phi}{\sqrt{2\pi \cdot 2Dt}} \exp\left[-\frac{(x - R_p)^2}{2 \cdot 2Dt}\right] \tag{14}$$

t_0 can then be evaluated

$$t_0 = \frac{\Delta R_p^2}{2D} \tag{15}$$

Therefore, the diffusion profile is then expressed by

$$N(x, t) = \frac{\Phi}{2\sqrt{\pi D(t + t_0)}} \exp\left[-\frac{(x - R_p)^2}{4D(t + t_0)}\right]$$

$$= \frac{\Phi}{2\sqrt{\pi D\left(t + \dfrac{\Delta R_p^2}{2D}\right)}} \exp\left[-\frac{(x - R_p)^2}{4D\left(t + \dfrac{\Delta R_p^2}{2D}\right)}\right]$$

$$= \frac{\Phi}{\sqrt{2\pi(\Delta R_p^2 + 2Dt)}} \exp\left[-\frac{(x - R_p)^2}{2(\Delta R_p^2 + 2Dt)}\right] \tag{16}$$

This is the same as Eq. 10.

In the case of various thermal processes, we can replace $\sqrt{\Delta R_p^2 + 2Dt}$ by $\sqrt{\Delta R_p^2 + 2\sum_i D_i t_i}$, where D_i and t_i are diffusion coefficient and time for the i-th thermal process, respectively.

Equation 7 was derived under the implicit assumption of infinite plane. In real, we have surface, and Eq. 10 is valid only when the diffusion occurs in the deep bulk where the surface concentration can be neglected.

We can easily extend the model to accommodate influence of surface if we assume the flux of the surface is zero or infinite.

Zero surface flux

One special boundary condition is zero flux at the surface. We tentatively modify the real initial condition of $f(x)$ by adding a symmetrical profile in the negative region given by

$$N(x, 0) = g(x) = \begin{cases} f(x) & \text{for } x \geq 0 \\ f(-x) & \text{for } x < 0 \end{cases} \tag{17}$$

Regarding this modified initial condition, the boundary condition of

$$\left. \frac{\partial N}{\partial x} \right|_{x=0} = 0 \tag{18}$$

is automatically satisfied. The corresponding diffusion profile is derived as

$$N(x, t) = \frac{1}{\sqrt{2\pi}} \int_{-\infty}^{\infty} g(\xi) \frac{1}{\sqrt{2Dt}} \exp\left(-\frac{(x-\xi)^2}{4Dt}\right) d\xi$$

$$= \frac{1}{\sqrt{2\pi}} \left[\int_{-\infty}^{0} f(-\xi) \frac{1}{\sqrt{2Dt}} \exp\left(-\frac{(x-\xi)^2}{4Dt}\right) d\xi + \int_{0}^{\infty} f(\xi) \frac{1}{\sqrt{2Dt}} \exp\left(-\frac{(x-\xi)^2}{4Dt}\right) d\xi \right]$$

$$= \frac{1}{2\sqrt{\pi Dt}} \int_{0}^{\infty} f(\xi) \left[\exp\left(-\frac{(x-\xi)^2}{4Dt}\right) + \exp\left(-\frac{(x+\xi)^2}{4Dt}\right) \right] d\xi \tag{19}$$

Fig. 1 shows the diffusion profiles. The gradient of the profile near the surface becomes flat with time.

We assume that the implanted profile $f(x)$ is Gaussian and $f(0)$ can be regarded as zero. We perform the integration and obtain an analytical form of

$$N(x, t) = \frac{1}{2\sqrt{\pi Dt}} \int_{-\infty}^{\infty} f(\xi) \left[\exp\left(-\frac{(x-\xi)^2}{4Dt}\right) + \exp\left(-\frac{(x+\xi)^2}{4Dt}\right) \right] d\xi$$

$$= \frac{\Phi}{\sqrt{2\pi(\Delta R_p^2 + 2Dt)}} \left[\exp\left(-\frac{(x-R_p)^2}{2(\Delta R_p^2 + 2Dt)}\right) + \exp\left(-\frac{(x+R_p)^2}{2(\Delta R_p^2 + 2Dt)}\right) \right] \tag{20}$$

The second term expresses that impurities are reflected at the surface and diffuse into the substrate instead of diffusion in the negative plane.

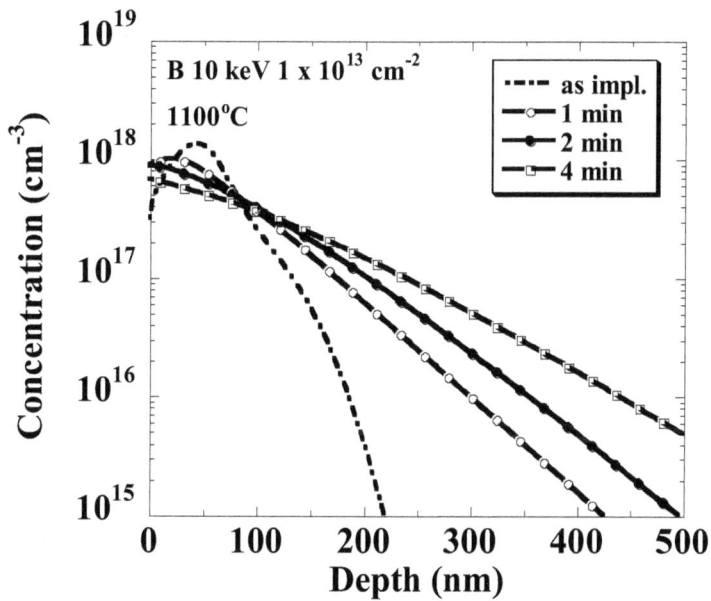

Figure 1: Analytical B diffusion profile with zero flux at the surface.

Infinite surface flux

The other special surface boundary condition is infinite surface flux condition. The corresponding boundary condition is expressed by

$$N(0,t) = 0 \tag{21}$$

We can express the boundary condition as

$$N(x, 0) = g(x) = \begin{cases} f(x) & for \ x \geq 0 \\ -f(-x) & for \ x < 0 \end{cases} \tag{22}$$

The corresponding solution is given by

$$N(x, t) = \frac{1}{2\sqrt{\pi Dt}} \int_0^\infty f(\xi)\left[\exp\left(-\frac{(x-\xi)^2}{4Dt}\right) - \exp\left(-\frac{(x+\xi)^2}{4Dt}\right)\right]d\xi \tag{23}$$

Fig. **2** shows the B diffusion profiles with this boundary condition. The concentration at the surface is zero as is expected. In this case, significant out diffusion occurs.

Assuming $f(x)$ of Gaussian profile, we perform the integration and obtain an analytical form as

$$N(x, t) = \frac{\Phi}{\sqrt{2\pi(\Delta R_p^2 + 2Dt)}}\left[\exp\left(-\frac{(x-R_p)^2}{2(\Delta R_p^2 + 2Dt)}\right) - \exp\left(-\frac{(x+R_p)^2}{2(\Delta R_p^2 + 2Dt)}\right)\right] \tag{24}$$

The second term corresponds to the out diffusion.

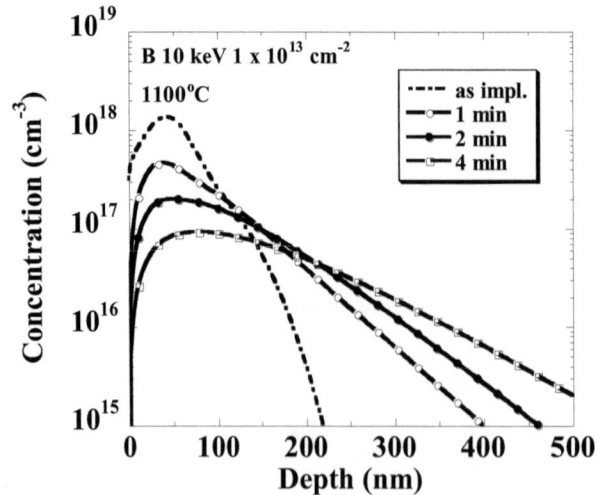

Figure 2: Diffusion profiles with infinity flux at the surface.

DEPENDENCE OF MOMENTS ON DIFFUSION TIME

We obtained explicit analytical form with initial Gaussian profile. The integration in Eq. 7 can be performed in special cases, and we cannot obtain it in general cases. However, we can obtain explicit model for moment parameters as shown below.

The diffusion profiles with an initial profile of $f(x)$ is given by

$$N(x,t) = \frac{1}{\sqrt{2\pi}} \int_{-\infty}^{\infty} f(\xi) \frac{1}{\sqrt{2Dt}} \exp\left[-\frac{(x-\xi)^2}{4Dt} \right] d\xi$$

$$= \frac{1}{\sqrt{\pi}L_D} \int_{-\infty}^{\infty} f(\xi) \exp\left[-\left(\frac{x-\xi}{L_D} \right)^2 \right] d\xi \tag{25}$$

where the $f(x)$ is normalized with respect to dose, where the diffusion length L_D is given by

$$L_D = 2\sqrt{Dt} \tag{26}$$

The dependence of R_p on diffusion time is evaluated as

$$R_p(t) = \int_{-\infty}^{\infty} x N(x,t) dx$$

$$= \frac{1}{\sqrt{\pi}L_D} \int_{-\infty}^{\infty} x \int_{-\infty}^{\infty} f(\xi) \exp\left[-\left(\frac{x-\xi}{L_D} \right)^2 \right] d\xi dx \tag{27}$$

$$= \frac{1}{\sqrt{\pi}L_D} \int_{-\infty}^{\infty} f(\xi) \int_{-\infty}^{\infty} x \exp\left[-\left(\frac{x-\xi}{L_D} \right)^2 \right] dx d\xi$$

Performing the integration with respect to x, we obtain

$$\int_{-\infty}^{\infty} x \exp\left(-\left(\frac{x-\xi}{L_D}\right)^2\right) dx = \int_{-\infty}^{\infty} \left[(x-\xi)+\xi\right] \exp\left[-\left(\frac{x-\xi}{L_D}\right)^2\right] dx$$

$$= L_D \int_{-\infty}^{\infty} \left(\frac{x-\xi}{L_D}\right) \exp\left[-\left(\frac{x-\xi}{L_D}\right)^2\right] dx$$

$$+ \xi \int_{-\infty}^{\infty} \exp\left[-\left(\frac{x-\xi}{L_D}\right)^2\right] dx \tag{28}$$

$$= \xi L_D \int_{-\infty}^{\infty} \exp\left(-s^2\right) dx$$

$$= \xi \sqrt{\pi} L_D$$

where

$$s = \frac{x-\xi}{L_D} \tag{29}$$

Substituting Eq. 28 into 27, we obtain

$$R_p(t) = \frac{1}{\sqrt{\pi}L_D} \int_{-\infty}^{\infty} f(\xi)\xi\sqrt{\pi}L_D d\xi$$

$$= \frac{1}{\sqrt{\pi}L_D} \int_{-\infty}^{\infty} f(\xi)\xi\sqrt{\pi}L_D d\xi \tag{30}$$

$$= \int_{-\infty}^{\infty} \xi f(\xi) d\xi$$

$$= R_p$$

It is noted that R_p is independent of diffusion time.

Next, we evaluate the second moment as

$$\Delta R_p^2(t) = \int_{-\infty}^{\infty} (x - R_p)^2 N(x,t)\,dx$$

$$= \frac{1}{\sqrt{\pi} L_D} \int_{-\infty}^{\infty} (x - R_p)^2 \int_{-\infty}^{\infty} f(\xi) \exp\left[-\left(\frac{x - \xi}{L_D}\right)^2\right] d\xi\,dx \qquad (31)$$

$$= \frac{1}{\sqrt{\pi} L_D} \int_{-\infty}^{\infty} f(\xi) \int_{-\infty}^{\infty} (x - R_p)^2 \exp\left[-\left(\frac{x - \xi}{L_D}\right)^2\right] dx\,d\xi$$

The integration with respect to x can be performed as

$$\int_{-\infty}^{\infty} (x - R_p)^2 \exp\left[-\left(\frac{x - \xi}{L_D}\right)^2\right] dx$$

$$= \int_{-\infty}^{\infty} \left[(x - \xi) + (\xi - R_p)\right]^2 \exp\left[-\left(\frac{x - \xi}{L_D}\right)^2\right] dx$$

$$= L_D^2 \int_{-\infty}^{\infty} \left(\frac{x - \xi}{L_D}\right)^2 \exp\left[-\left(\frac{x - \xi}{L_D}\right)^2\right] dx$$

$$+ 2L_D(\xi - R_p) \int_{-\infty}^{\infty} \frac{x - \xi}{L_D} \exp\left[-\left(\frac{x - \xi}{L_D}\right)^2\right] dx$$

$$+ (\xi - R_p)^2 \int_{-\infty}^{\infty} \exp\left[-\left(\frac{x - \xi}{L_D}\right)^2\right] dx$$

$$= L_D^3 \int_{-\infty}^{\infty} s^2 \exp(-s^2)\,dx$$

$$+ 2L_D^2(\xi - R_p) \int_{-\infty}^{\infty} s \exp(-s^2)\,dx$$

$$\vdots\; L_D(\xi - R_p)^2 \int_{-\infty}^{\infty} \exp(-s^2)\,dx$$

$$= \frac{L_D^3}{2}\sqrt{\pi} + L_D(\xi - R_p)^2 \sqrt{\pi} \qquad (32)$$

Substituting Eq. 32 into Eq. 31, we obtain

$$\Delta R_p^{\;2}(t) = \frac{1}{\sqrt{\pi}L_D} \int_{-\infty}^{\infty} f(\xi) \left[\frac{L_D^{\;3}}{2} \sqrt{\pi} + L_D (\xi - R_p)^2 \sqrt{\pi} \right] d\xi$$

$$= \frac{1}{\sqrt{\pi}L_D} \frac{L_D^{\;3}}{2} \sqrt{\pi} + \frac{1}{\sqrt{\pi}L_D} L_D \sqrt{\pi} \int_{-\infty}^{\infty} f(\xi) L(\xi - R_p)^2 d\xi \qquad (33)$$

$$= \frac{L_D^{\;2}}{2} + \int_{-\infty}^{\infty} f(\xi)(\xi - R_p)^2 d\xi$$

$$= 2Dt + \Delta R_p^{\;2}$$

ΔR_p monotonically increases with increasing diffusion time.

Evaluating the third order moment μ_3, we obtain

$$\mu_3(t) = \int_{-\infty}^{\infty} (x - R_p)^3 N(x,t) dx$$

$$= \frac{1}{\sqrt{\pi}L_D} \int_{-\infty}^{\infty} (x - R_p)^3 \int_{-\infty}^{\infty} f(\xi) \exp\left(-\left(\frac{x - \xi}{L_D} \right)^2 \right) d\xi dx \qquad (34)$$

$$= \frac{1}{\sqrt{\pi}L_D} \int_{-\infty}^{\infty} f(\xi) \int_{-\infty}^{\infty} (x - R_p)^3 \exp\left(-\left(\frac{x - \xi}{L_D} \right)^2 \right) dx d\xi$$

The integration with respect to x can be performed as

$$\int_{-\infty}^{\infty} \left(x - R_p\right)^3 \exp\left[-\left(\frac{x-\xi}{L_D}\right)^2\right] dx$$

$$= \int_{-\infty}^{\infty} \left[(x-\xi) + \left(\xi - R_p\right)\right]^3 \exp\left[-\left(\frac{x-\xi}{L_D}\right)^2\right] dx$$

$$= L_D^{\ 3} \int_{-\infty}^{\infty} \left(\frac{x-\xi}{L_D}\right)^3 \exp\left[-\left(\frac{x-\xi}{L_D}\right)^2\right] dx$$

$$+ 3L_D^{\ 2}\left(\xi - R_p\right) \int_{-\infty}^{\infty} \left(\frac{x-\xi}{L_D}\right)^2 \exp\left[-\left(\frac{x-\xi}{L_D}\right)^2\right] dx$$

$$+ 3L_D\left(\xi - R_p\right)^2 \int_{-\infty}^{\infty} \left(\frac{x-\xi}{L_D}\right) \exp\left[-\left(\frac{x-\xi}{L_D}\right)^2\right] dx$$

$$+ \left(\xi - R_p\right)^3 \int_{-\infty}^{\infty} \exp\left[-\left(\frac{x-\xi}{L_D}\right)^2\right] dx$$

$$= 3L_D^{\ 3}\left(\xi - R_p\right) \int_{-\infty}^{\infty} s^2 \exp\left(-s^2\right) dx + L_D\left(\xi - R_p\right)^3 \int_{-\infty}^{\infty} \exp\left(-s^2\right) dx \qquad (35)$$

$$= \frac{3\sqrt{\pi}}{2} L_D^{\ 3}\left(\xi - R_p\right) + \sqrt{\pi} L_D\left(\xi - R_p\right)^3$$

Substituting Eq. 35 into Eq. 34, we obtain

$$\mu_3(t) = \frac{1}{\sqrt{\pi}L_D} \int_{-\infty}^{\infty} f(\xi)\left[\frac{3\sqrt{\pi}}{2}L_D^{\ 3}\left(\xi - R_p\right) + \sqrt{\pi}L_D\left(\xi - R_p\right)^3\right] d\xi$$

$$= \frac{1}{\sqrt{\pi}L_D}\sqrt{\pi}L_D \int_{-\infty}^{\infty} \left(\xi - R_p\right)^3 f(\xi) d\xi \qquad (36)$$

$$= \mu_3$$

Consequently, the third moment is independent of diffusion time. The dependence of γ, which is defined below, can then be evaluated as

$$\gamma(t) = \frac{\mu_3(t)}{\Delta R_p(t)^3}$$

$$= \frac{\mu_3}{\left(\Delta R_p^2 + 2Dt\right)^{\frac{3}{2}}}$$

$$= \gamma \left(\frac{\Delta R_p^2}{\Delta R_p^2 + 2Dt}\right)^{\frac{3}{2}} \tag{37}$$

Therefore, the profile becomes symmetrical with time.

The forth moment μ_4 can be evaluated as

$$\mu_4(t) = \int_{-\infty}^{\infty} (x - R_p)^4 N(x,t) dx$$

$$= \frac{1}{\sqrt{\pi} L_D} \int_{-\infty}^{\infty} (x - R_p)^4 \int_{-\infty}^{\infty} f(\xi) \exp\left(-\left(\frac{x-\xi}{L_D}\right)^2\right) d\xi dx \tag{38}$$

$$= \frac{1}{\sqrt{\pi} L_D} \int_{-\infty}^{\infty} f(\xi) \int_{-\infty}^{\infty} (x - R_p)^4 \exp\left(-\left(\frac{x-\xi}{L_D}\right)^2\right) dx d\xi$$

The integration with respect to x can be performed as

$$\int_{-\infty}^{\infty} \left(x - R_p\right)^4 \exp\left(-\left(\frac{x - \xi}{L_D}\right)^2\right) dx$$

$$= \int_{-\infty}^{\infty} \left[(x - \xi) + (\xi - R_p)\right]^4 \exp\left(-\left(\frac{x - \xi}{L_D}\right)^2\right) dx$$

$$= L_D{}^4 \int_{-\infty}^{\infty} \left(\frac{x - \xi}{L_D}\right)^4 \exp\left(-\left(\frac{x - \xi}{L_D}\right)^2\right) dx$$

$$+ 4L_D{}^3 \left(\xi - R_p\right) \int_{-\infty}^{\infty} \left(\frac{x - \xi}{L_D}\right)^3 \exp\left(-\left(\frac{x - \xi}{L_D}\right)^2\right) dx$$

$$+ 6L_D{}^2 \left(\xi - R_p\right)^2 \int_{-\infty}^{\infty} \left(\frac{x - \xi}{L_D}\right)^2 \exp\left(-\left(\frac{x - \xi}{L_D}\right)^2\right) dx$$

$$+ 4L_D \left(\xi - R_p\right)^3 \int_{-\infty}^{\infty} \left(\frac{x - \xi}{L_D}\right) \exp\left(-\left(\frac{x - \xi}{L_D}\right)^2\right) dx$$

$$+ \left(\xi - R_p\right)^4 \int_{-\infty}^{\infty} \exp\left(-\left(\frac{x - \xi}{L_D}\right)^2\right) dx$$

$$= L_D{}^5 \int_{-\infty}^{\infty} s^4 \exp\left(-s^2\right) dx + 6L_D{}^3 \left(\xi - R_p\right)^2 \int_{-\infty}^{\infty} s^2 \exp\left(-s^2\right) dx \qquad (39)$$

$$+ L_D \left(\xi - R_p\right)^4 \int_{-\infty}^{\infty} \exp\left(-s^2\right) dx$$

$$= L_D{}^5 \frac{3\sqrt{\pi}}{4} + 6L_D{}^3 \left(\xi - R_p\right)^2 \frac{\sqrt{\pi}}{2} + L_D \left(\xi - R_p\right)^4 \sqrt{\pi}$$

Substituting Eq. 39 into Eq. 38, we obtain

$$\mu_4(t) = \frac{1}{\sqrt{\pi}L_D} \int_{-\infty}^{\infty} f(\xi) \left[L_D{}^5 \frac{3\sqrt{\pi}}{4} + 6L_D{}^3 \left(\xi - R_p\right)^2 \frac{\sqrt{\pi}}{2} + L_D \left(\xi - R_p\right)^4 \sqrt{\pi} \right] d\xi$$

$$= \frac{1}{\sqrt{\pi}L_D} L_D{}^5 \frac{3\sqrt{\pi}}{4}$$

$$+ \frac{1}{\sqrt{\pi}L_D} 3L_D{}^3 \sqrt{\pi} \int_{-\infty}^{\infty} \left(\xi - R_p\right)^2 f(\xi) d\xi \qquad (40)$$

$$+ \frac{1}{\sqrt{\pi}L_D} L_D \sqrt{\pi} \int_{-\infty}^{\infty} \left(\xi - R_p\right)^4 f(\xi) d\xi$$

$$= 12\left(Dt\right)^2 + 12Dt\Delta R_p{}^2 + \mu_4$$

Therefore, β, which is defined below, is given by

$$\beta(t) = \frac{\mu_4(t)}{\Delta R_p(t)^4}$$

$$= \frac{12\left(Dt\right)^2 + 12Dt\Delta R_p{}^2 + \mu_4}{\left(\Delta R_p{}^2 + 2Dt\right)^2}$$

$$= \frac{3\left(\dfrac{2Dt}{\Delta R_p{}^2}\right)^2 + 6\dfrac{2Dt}{\Delta R_p{}^2} + \beta}{\left(1 + \dfrac{2Dt}{\Delta R_p{}^2}\right)^2}$$

$$(41)$$

Fig. **3** compares the analytical models for time dependent moments with numerical ones. The analytical models agree well with numerical ones. R_p is invariant and ΔR_p monotonically increases with diffusion time as are expected. γ approaches to zero, and β to 3 with increasing diffusion time. The dependence of γ is clear from Eq. 37.

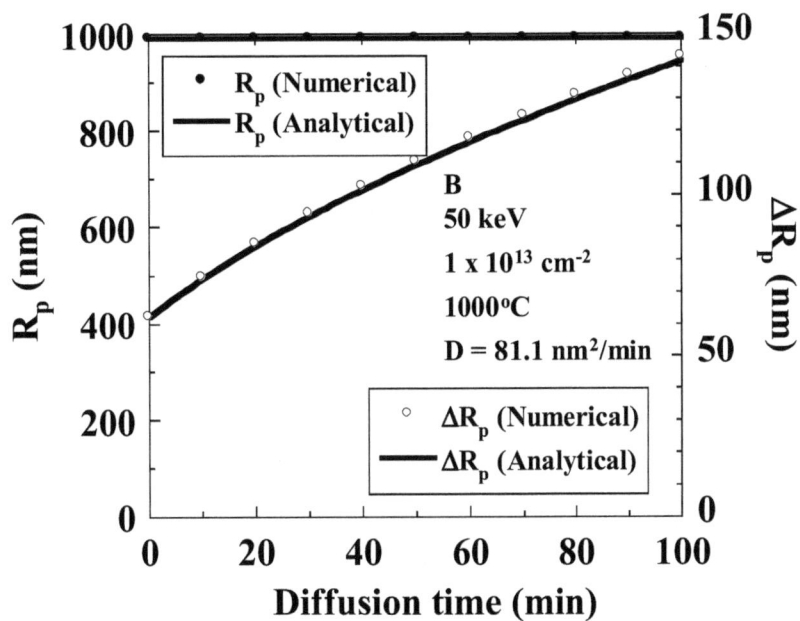

(a)

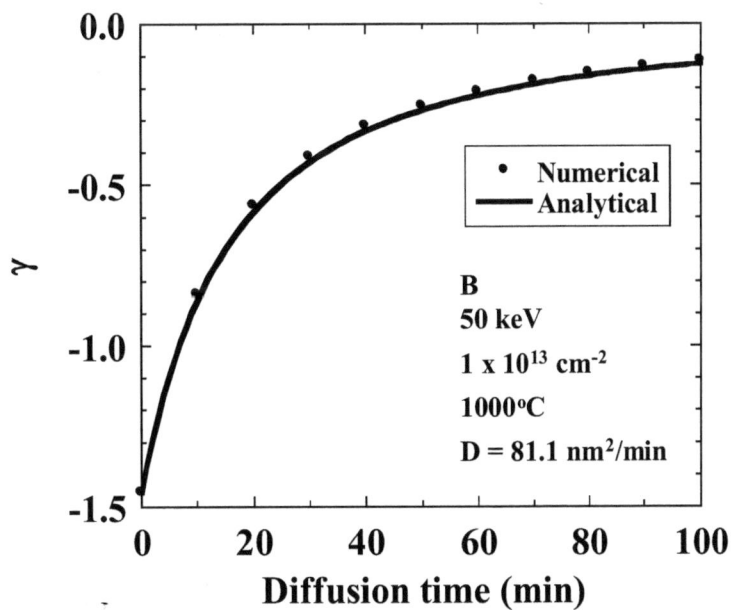

(b)

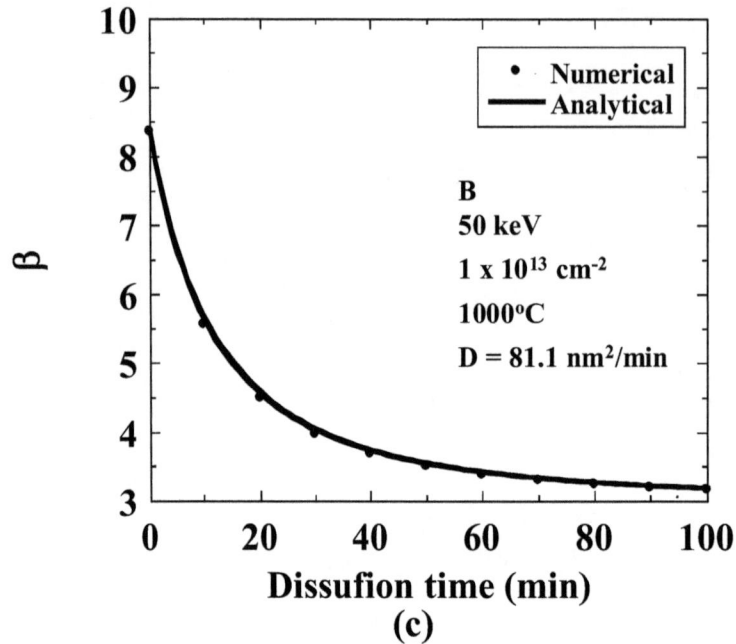

Figure 3: Dependence of moment parameters on diffusion time.

When time goes on so that $2Dt \gg \Delta R_p^{\,2}$ becomes hold, $\beta(t)$ is reduced to

$$
\begin{aligned}
\beta(t) &= \frac{3\left(\dfrac{2Dt}{\Delta R_p^{\,2}}\right)^2 + 6\dfrac{2Dt}{\Delta R_p^{\,2}} + \beta}{\left(1 + \dfrac{2Dt}{\Delta R_p^{\,2}}\right)^2} \\[2em]
&\approx \frac{3\left(\dfrac{2Dt}{\Delta R_p^{\,2}}\right)^2}{\left(\dfrac{2Dt}{\Delta R_p^{\,2}}\right)^2} \\[2em]
&= 3
\end{aligned}
\tag{42}
$$

Therefore, when the diffusion proceeds, any profile becomes Gaussian profiles.

Let us consider that an initial profile is Gaussian, where $\gamma = 0$ and $\beta = 3$. In this special case,

$$\gamma(t) = 0 \times \left(\frac{\Delta R_p^{\,2}}{\Delta R_p^{\,2} + 2Dt} \right)^{\frac{3}{2}} = 0 \tag{43}$$

$$\beta(t) = \frac{3\left(\dfrac{2Dt}{\Delta R_p^{\,2}}\right)^2 + 6\dfrac{2Dt}{\Delta R_p^{\,2}} + 3}{\left(1 + \dfrac{2Dt}{\Delta R_p^{\,2}}\right)^2}$$

$$= 3\frac{\left(\dfrac{2Dt}{\Delta R_p^{\,2}}\right)^2 + 2\dfrac{2Dt}{\Delta R_p^{\,2}} + 1}{\left(1 + \dfrac{2Dt}{\Delta R_p^{\,2}}\right)^2}$$

$$= 3\frac{\left(\dfrac{2Dt}{\Delta R_p^{\,2}} + 1\right)^2}{\left(1 + \dfrac{2Dt}{\Delta R_p^{\,2}}\right)^2}$$

$$= 3 \tag{44}$$

Consequently, $\gamma = 0$ and $\beta = 3$ holds over entire diffusion time with Gaussian profiles.

PROFILES WITH CONSTANT CONCENTRATION DIFFUSION SOURCE

The solid diffusion source or vapor phase diffusion source provide the constant concentration diffusion situation. Impurities form clusters in high concentration region, and the clusters play a role of constant concentration diffusion source. Let us see what profiles we can expect with this situation.

The diffusion equation is given by

$$\frac{\partial N(x,t)}{\partial t} = D\frac{\partial^2(x,t)}{\partial x^2}$$

(45)

with an initial condition of

$$N(x,0) = 0$$

(46)

The boundary conditions are given by

$$\begin{cases} N(0,t) = N_0 \\ N(\infty,t) = 0 \end{cases}$$

(47)

Performing Laplace transformation of Eq. 45 with respect to time, we obtain

$$s\mathscr{L}\{N(x,t)\} = D\frac{d^2\mathscr{L}\{N(x,t)\}}{dx^2}$$

(48)

This can be solved as

$$\mathscr{L}\{N(x,t)\} = A(s)\exp\left(-\sqrt{\frac{s}{D}}x\right)$$

(49)

Performing the Laplace transformation to the boundary condition, we obtain

$$\mathscr{L}\{N(0,t)\} = A(s) = \frac{N_0}{s}$$

(50)

Substituting Eq. 50 into Eq. 49, we obtain

$$\mathscr{L}\{N(x,t)\} = \frac{N_0}{s}\exp\left(-\sqrt{\frac{s}{D}}x\right)$$

(51)

Performing inverse Laplace transformation, we obtain

$$N(x,t) = \mathscr{L}^{-1}\left\{\frac{N_0}{s}\exp\left(-\sqrt{\frac{s}{D}}x\right)\right\}$$

$$= N_0 erfc\left(\frac{x}{2\sqrt{Dt}}\right)$$

(52)

Fig. **4** shows the diffusion profiles evaluated with Eq. 52.

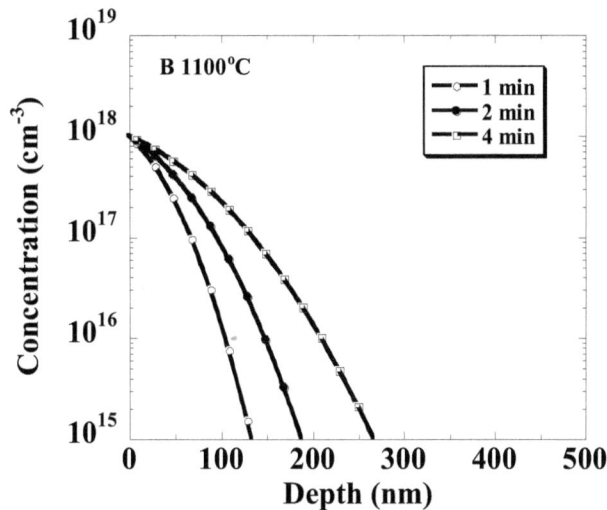

Figure 4: Diffusion profiles with constant concentration diffusion source.

DIFFUSION PROFILES WITH CONCENTRATION DEPENDENT DIFFUSION COEFFICIENT

When impurity concentration N exceeds intrinsic carrier concentration n_i, the dependence of diffusion coefficient on the concentration becomes significant and the assumption of constant diffusion coefficient becomes invalid. The diffusion equation then becomes nonlinear, and the derivation of corresponding models becomes difficult to solve.

Nakajima and Fair proposed empirical analytical models [3-6]. These models well reproduced experimental data. However, these are mathematical models which are not based on diffusion equation, and hence lack productive value. Anderson succeeded in deriving a model with constant dose of impurity based on nonlinear diffusion equation [7, 8]. This Anderson's model was refined by introducing a normalized distance, and further a model with constant concentration diffusion source was derived [9-11]. The analytical model are compared with numerical data calculated using a one-dimensional process simulator TESIM [12].

NORMALIZED DIFFUSION EQUATION IN HIGH CONCENTRATION REGION

The diffusion equation in high concentration region is given by

$$\frac{\partial N}{\partial t} = \frac{\partial}{\partial x}\left(hD\frac{\partial N}{\partial x}\right) \tag{53}$$

where h expresses drift effect associated with electric field and is given by

$$h = 1 + \frac{1}{\sqrt{1 + \left(\frac{2n_i}{N}\right)^2}} \tag{54}$$

D is the diffusion coefficient and its general form is given by

$$D = D_i^x + \frac{n}{n_i}D_i^m + \left(\frac{n}{n_i}\right)^2 D_i^{mm} + \frac{p}{n_i}D_i^p \tag{55}$$

D_i^x, D_i^m, D_i^{mm}, and D_i^p are the diffusion coefficients associated with neutral, single negative, double negative, and single positive charged point defect impurity pair, respectively. n is the electron concentration, p is the whole concentration. In general, donor or acceptor dominate, and both are expressed by impurity concentration N with

$$n,p = \frac{N}{2} + \sqrt{\left(\frac{N}{2}\right)^2 + n_i^2} \tag{56}$$

Assuming $N \gg n_i$ and we can approximate as follows.

$$h \approx 2 \tag{57}$$
$$n, p \approx N \tag{58}$$

We further assume that only one term in Eq. 55 dominate D and express an approximate diffusion coefficient of Eq. 55 as

$$D = \left(\frac{n}{n_i}\right)^\gamma D_i \tag{59}$$

If we want to empirically express some terms in Eq. 55 using the form of Eq. 59, we regard this γ is effective one and hence is not necessary to be integer. The diffusion equation is given by

$$\frac{\partial N}{\partial t} = \frac{\partial}{\partial x}\left[2\left(\frac{n}{n_i}\right)^{\gamma} D_i \frac{\partial N}{\partial x}\right]$$

(60)

Introducing a variable of

$$\frac{N}{n_i} \equiv s$$

(61)

we obtain diffusion equation as

$$\frac{\partial s}{\partial t} = 2D_i \frac{\partial}{\partial x}\left(s^{\gamma} \frac{\partial s}{\partial x}\right)$$

(62)

Anderson further introduced following variables [7].

$$\xi = \frac{x}{(2D_i t)^k}$$

(63)

$$\tau = 2D_i t$$

(64)

The left side of the Eq. 62 is performed as

$$\frac{\partial s}{\partial t} = \frac{\partial s}{\partial \xi}\frac{\partial \xi}{\partial t} + \frac{\partial s}{\partial \tau}\frac{\partial \tau}{\partial t}$$
$$= -k\frac{\xi}{t}\frac{\partial s}{\partial \xi} + 2D_i\frac{\partial s}{\partial \tau}$$
$$= -2D_i k\frac{\xi}{\tau}\frac{\partial s}{\partial \xi} + 2D_i\frac{\partial s}{\partial \tau}$$

The right side of the Eq. 62 is performed as

$$2D_i\frac{\partial}{\partial x}\left(s^{\gamma}\frac{\partial s}{\partial x}\right) = 2D_i\frac{1}{(2D_i t)^{2k}}\frac{\partial}{\partial \xi}\left(s^{\gamma}\frac{\partial s}{\partial \xi}\right)$$
$$= 2D_i\frac{1}{\tau^{2k}}\frac{\partial}{\partial \xi}\left(s^{\gamma}\frac{\partial s}{\partial \xi}\right)$$

We then obtain diffusion equation given by

$$\frac{1}{\tau^{2k}}\frac{\partial}{\partial \xi}\left(s^{\gamma}\frac{\partial s}{\partial \xi}\right) + k\frac{\xi}{\tau}\frac{\partial s}{\partial \xi} = \frac{\partial s}{\partial \tau}$$

(65)

This is a basic diffusion equation to solve.

CONSTANT IMPURITY DOSE

We consider the situation in which impurity dose is constant at Q_T [7, 8]. We set an initial condition of delta doped one located in infinite plane at $x = 0$ $(\xi = 0)$ given by

$$N(x,0) = Q_T\delta(x) \tag{66}$$

We assume the solution of the equation with a form of

$$s = \tau^\alpha \phi(\xi) \tag{67}$$

Converting normalized variables to real variables, we obtain

$$N = n_i(2D_it)^\alpha \phi\left(\frac{x}{(2D_it)^k}\right) \tag{68}$$

Q_T is then given by

$$Q_T = 2\int_0^\infty N dx$$
$$= 2n_i\tau^{\alpha+k}\int_0^\xi \phi(\xi)d\xi \tag{69}$$

Since Q_T is constant, it does not depend on τ. We hence obtain

$$\alpha + k = 0 \tag{70}$$

Substituting Eq. 67 into Eq. 65, we obtain

$$\frac{1}{\tau^{2k}}\frac{\partial}{\partial\xi}\left(\tau^{\alpha\gamma}\phi^\gamma\frac{\partial}{\partial\xi}(\tau^\alpha\phi)\right) + k\frac{\xi}{\tau}\frac{\partial}{\partial\xi}(\tau^\alpha\phi) = \alpha\tau^{\alpha-1}\phi$$

This can be arranged as

$$\tau^{\alpha\gamma+\alpha-2k}\frac{\partial}{\partial\xi}\left(\phi^\gamma\frac{\partial\phi}{\partial\xi}\right) + \tau^{\alpha-1}k\xi\frac{\partial\phi}{\partial\xi} = \alpha\tau^{\alpha-1}\phi \tag{71}$$

We impose that this is independent of τ, and obtain

$$\alpha\gamma + \alpha - 2k = \alpha - 1 \tag{72}$$

From Eqs. 70 and 72, we obtain

$$k = -\alpha = \frac{1}{\gamma + 2} \tag{73}$$

Equation 71 is then reduced to

$$\frac{\partial}{\partial\xi}\left(\phi^{\gamma}\frac{\partial\phi}{\partial\xi}\right) + \frac{1}{\gamma+2}\left(\xi\frac{\partial\phi}{\partial\xi} + \phi\right) = 0 \tag{74}$$

Utilizing the relationship of

$$\frac{\partial(\xi\phi)}{\partial\xi} = \xi\frac{\partial\phi}{\partial\xi} + \phi \tag{75}$$

we further reduce Eq. 74 to

$$\frac{\partial}{\partial\xi}\left(\phi^{\gamma}\frac{\partial\phi}{\partial\xi} + \frac{1}{\gamma+2}\xi\phi\right) = 0 \tag{76}$$

This can be solved as

$$\phi^{\gamma}\frac{\partial\phi}{\partial\xi} + \frac{1}{\gamma+2}\xi\phi = A \tag{77}$$

From the symmetry of the system, we can assume that $\partial\Phi/\partial\xi=0$ for $\xi=0$. We then obtain $A=0$. Equation 77 is then reduced to

$$\phi^{\gamma-1}\frac{\partial\phi}{\partial\xi} + \frac{1}{\gamma+2}\xi = 0 \tag{78}$$

We denote ϕ for $\xi=0$ as ϕ_0, and integrate Eq. 78 with respect to ξ from 0 to ξ and with respect to ϕ from ϕ_0 to ϕ as

$$\int_{\phi_0}^{\phi}\phi^{\gamma-1}d\phi + \frac{1}{\gamma+2}\int_0^{\xi}\xi d\xi = \frac{1}{\gamma}\left[\phi^{\gamma} - \phi_0^{\gamma}\right] + \frac{1}{2(\gamma+2)}\xi^2 = 0$$

We then obtain

$$\phi(\xi) = \phi_0 \left[1 - \left(\frac{\xi}{\xi_0} \right)^2 \right]^{\frac{1}{\gamma}}$$

(79)

where

$$\xi_0{}^2 = \frac{2(\gamma + 2)}{\gamma} \phi_0{}^\gamma$$

(80)

ϕ_0 can be related to Q_T as

$$Q_T = 2 n_i \int_0^{\xi_0} \phi_0 \left[1 - \left(\frac{\xi}{\xi_0} \right)^2 \right]^{\frac{1}{\gamma}} d\xi$$

$$= 2 n_i \xi_0 \phi_0 \int_0^1 \left(1 - u^2 \right)^{\frac{1}{\gamma}} du$$

$$= 2 n_i \sqrt{\frac{2(\gamma + 2)}{\gamma} \phi_0{}^{\gamma+2}} \, I_\gamma$$

This can be solved with respect to ϕ_0 as

$$\phi_0 = \left[\frac{\gamma}{2(\gamma + 2)} \left(\frac{Q_T}{2 n_i I_\gamma} \right)^2 \right]^{\frac{1}{\gamma+2}}$$

(81)

where I_γ is

$$I_\gamma = \int_0^1 \left(1 - u^2 \right)^{\frac{1}{\gamma}} du$$

(82)

We further evaluate the integration of Eq. 82. Introducing a variable as $u^2 = l$, we obtain an analytical form given by

$$I_\gamma = \int_0^1 \frac{1}{2u}(1-l)^{\frac{1}{\gamma}}dl$$

$$= \frac{1}{2}\int_0^1 l^{-\frac{1}{2}}(1-l)^{\frac{1}{\gamma}}dl$$

$$= \frac{1}{2}\int_0^1 l^{\frac{1}{2}-1}(1-l)^{(\frac{1}{\gamma}+1)-1}dl$$

$$= \frac{1}{2}B\left(\frac{1}{2}, \frac{1}{\gamma}+1\right)$$

$$= \frac{1}{2}\frac{\Gamma\left(\frac{1}{2}\right)\Gamma\left(\frac{1}{\gamma}+1\right)}{\Gamma\left(\frac{3}{2}+\frac{1}{\gamma}\right)}$$

$$= \frac{\sqrt{\pi}}{2}\frac{\frac{1}{\gamma}\Gamma\left(\frac{1}{\gamma}\right)}{\left(\frac{1}{2}+\frac{1}{\gamma}\right)\Gamma\left(\frac{1}{2}+\frac{1}{\gamma}\right)}$$

$$= \frac{\sqrt{\pi}}{2+\gamma}\frac{\Gamma\left(\frac{1}{\gamma}\right)}{\Gamma\left(\frac{1}{2}+\frac{1}{\gamma}\right)} \tag{83}$$

where Γ is a Gamma function, and B is a Beta function.

Let us consider special cases of γ is 1 or 2. The values of Γ are given by

$$\Gamma\left(\frac{1}{2}\right) = \sqrt{\pi} \tag{84}$$

$$\Gamma(1) = \Gamma(2) = 1 \tag{85}$$

$$\Gamma\left(\frac{3}{2}\right) = \frac{1}{2}\Gamma\left(\frac{1}{2}\right) = \frac{\sqrt{\pi}}{2} \tag{86}$$

$$\Gamma\left(\frac{5}{2}\right) = \frac{3}{2}\Gamma\left(\frac{3}{2}\right) = \frac{3\sqrt{\pi}}{4} \tag{87}$$

Therefore, I_1, I_2 are given by

$$I_1 = \frac{1}{2}\frac{\Gamma\left(\frac{1}{2}\right)\Gamma(2)}{\Gamma\left(\frac{5}{2}\right)} = \frac{2}{3} \tag{88}$$

$$I_2 = \frac{1}{2} \frac{\Gamma\left(\frac{1}{2}\right)\Gamma\left(\frac{3}{2}\right)}{\Gamma(2)} = \frac{\pi}{4} \tag{89}$$

The peak concentration N_s can be evaluated by setting $x = 0$ in Eq. 68, and is given by

$$
\begin{aligned}
N_s &= n_i (2D_i t)^{-\frac{1}{\gamma+2}} \phi_0 \\
&= n_i (2D_i t)^{-\frac{1}{\gamma+2}} \left[\frac{\gamma}{2(\gamma+2)} \left(\frac{Q_T}{2n_i I_\gamma} \right)^2 \right]^{\frac{1}{\gamma+2}} \\
&= n_i \left[\frac{\gamma}{2(\gamma+2)} \left(\frac{Q_T}{2n_i I_\gamma} \right)^2 \frac{1}{2D_i t} \right]^{\frac{1}{\gamma+2}}
\end{aligned}
\tag{90}
$$

ξ_0 can be expressed with N_s as

$$
\begin{aligned}
\xi_0^{\,2} &= \frac{2(\gamma+2)}{\gamma} \left[\frac{N_s}{n_i} (2D_i t)^{\frac{1}{\gamma+2}} \right]^\gamma \\
&= \frac{2(\gamma+2)}{\gamma} \left(\frac{N_s}{n_i} \right)^\gamma (2D_i t)^{\frac{\gamma}{\gamma+2}}
\end{aligned}
\tag{91}
$$

Therefore, we obtain

$$
\begin{aligned}
\left(\frac{\xi}{\xi_0} \right)^2 &= \frac{1}{\dfrac{2(\gamma+2)}{\gamma} \left(\dfrac{N_s}{n_i} \right)^\gamma (2D_i t)^{\frac{\gamma}{\gamma+2}}} \frac{x^2}{(2D_i t)^{\frac{2}{\gamma+2}}} \\
&= \frac{x^2}{\dfrac{\gamma+2}{\gamma} \left(\dfrac{N_s}{n_i} \right)^\gamma 4D_i t}
\end{aligned}
\tag{92}
$$

Consequently, the diffusion profile is given by

$$N = N_s \left(1 - z_\gamma^{\,2} \right)^{\frac{1}{\gamma}} \tag{93}$$

where z_γ is a normalized distance defined by

$$z_\gamma^{\,2} \equiv \frac{x^2}{\dfrac{\gamma+2}{\gamma} \left(\dfrac{N_s}{n_i} \right)^\gamma 4D_i t} \tag{94}$$

Equation 93 becomes zero for $z_\gamma = 1$, and the gradient is quite high. We therefore, assume junction depth x_j is the x that satisfies $z_\gamma = 1$, that is

$$x_j = \sqrt{\frac{\gamma + 2}{\gamma}\left(\frac{N_s}{n_i}\right)^\gamma 4D_i t}$$

(95)

The time dependence of x_j can be clearly shown by substituting Eq. 90 of N_s given by

$$x_j = \sqrt{\frac{\gamma + 2}{\gamma}\left[\frac{\gamma}{\gamma + 2}\left(\frac{Q_T}{2n_iI_\gamma}\right)^2\frac{1}{4D_i t}\right]^{\frac{\gamma}{\gamma+2}} 4D_i t}$$

$$= \sqrt{\frac{\gamma + 2}{\gamma}\left[\frac{\gamma}{\gamma + 2}\left(\frac{Q_T}{2n_iI_\gamma}\right)^2\right]^{\frac{\gamma}{\gamma+2}}} (4D_i t)^{\frac{1}{\gamma+2}}$$

(96)

Fig. **5** shows the comparison of analytical and numerical results for B diffusion profile with $\gamma = 1$. The analytical model agrees well with numerical one. We use initial condition of 3×10^{20} cm^{-3} and 4 nm box profiles in the numerical simulation.

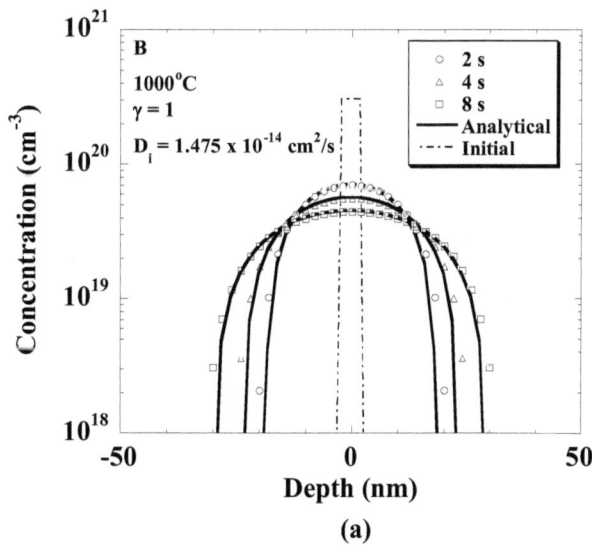

(a)

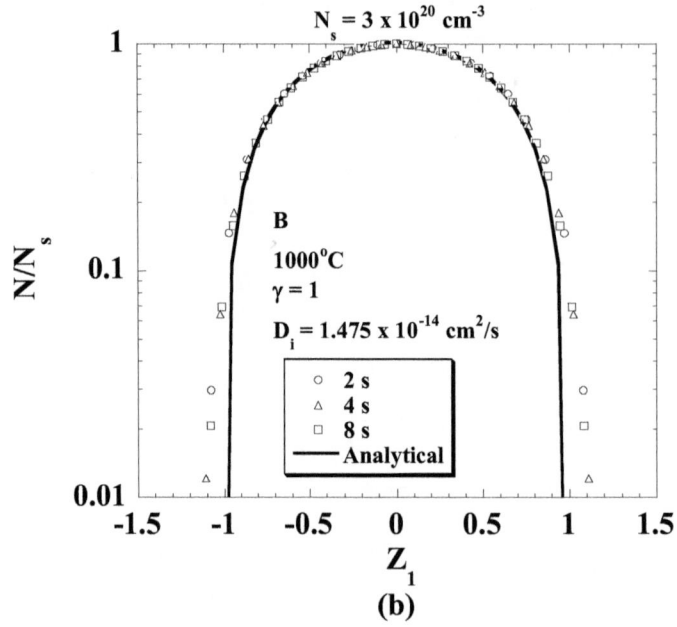

Figure 5: B diffusion profile ($\gamma = 1$) (a) Real distance (b) Normalized distance.

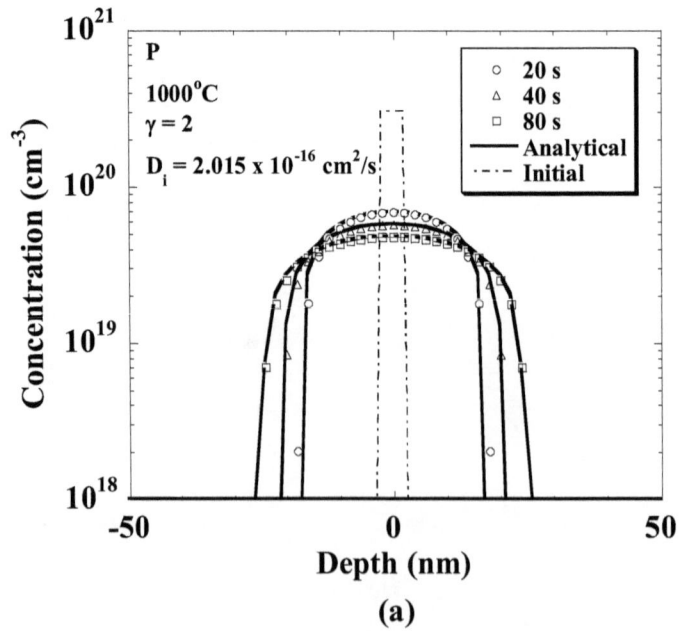

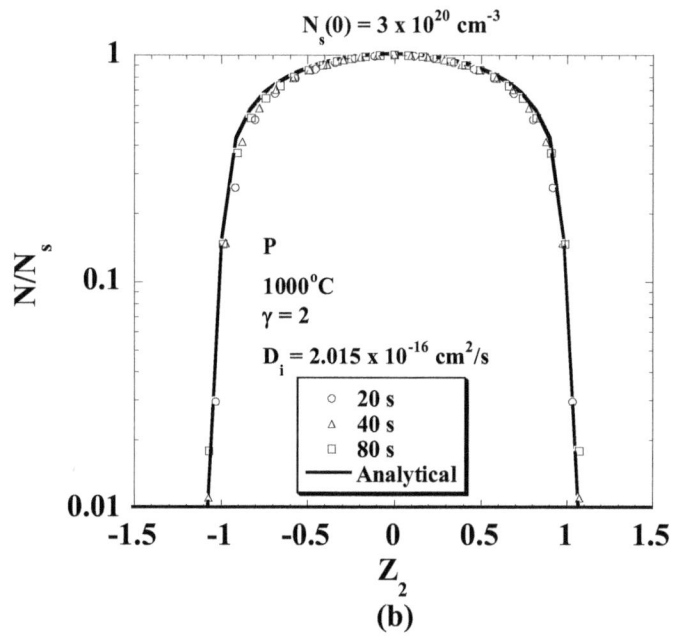

Figure 6: P diffusion profile ($\gamma = 2$) (a) Real distance (b) Normalized distance.

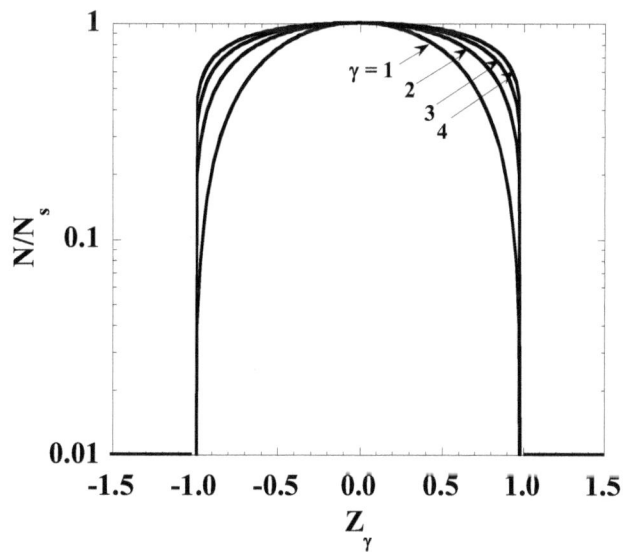

Figure 7: Dependence of normalized diffusion profile on γ.

Fig. **6** shows the comparison of analytical and numerical results for P diffusionprofiles with $\gamma = 2$. The analytical model agrees well with numerical one. The profile shapes are more box-like than those for B.

We can clearly compare shape of the profiles using normalized plot with Z_γ as shown in Fig. **7**. It is clear that the profile become box like with increasing γ.

Fig. **8** compares numerical and analytical results of the dependence of peak concentration on diffusion time. The analytical model agrees well with numerical one. This means that the peak concentration decreases in accordance with $t^{-\frac{1}{\gamma+2}}$.

We evaluated the numerical junction depth where the impurity concentration was $10^{17}\ cm^{-3}$. Fig. **9** compares dependence of junction depth x_j on diffusion time. The analytical model agrees well with numerical one. However, we observe deviation between them as increasing annealing time. As the diffusion time increases, the peak concentration decreases, and the assumption of $N \gg n_i$ into the analytical model becomes invalid for entire region.

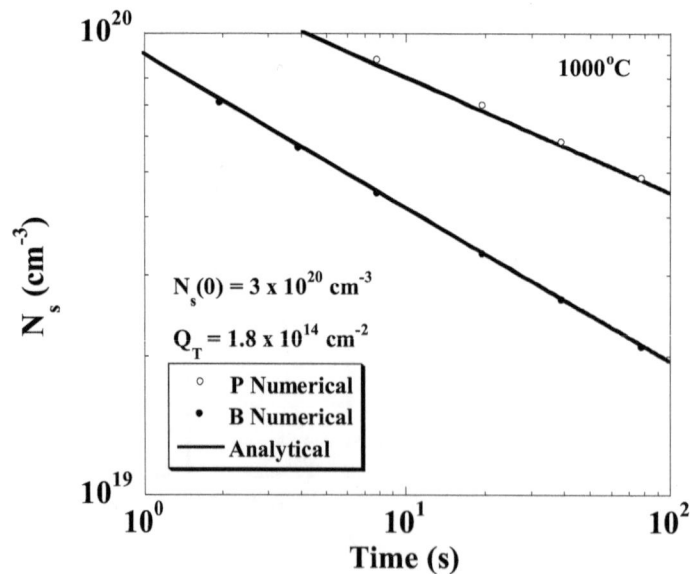

Figure 8: Dependence of peak concentration on diffusion time.

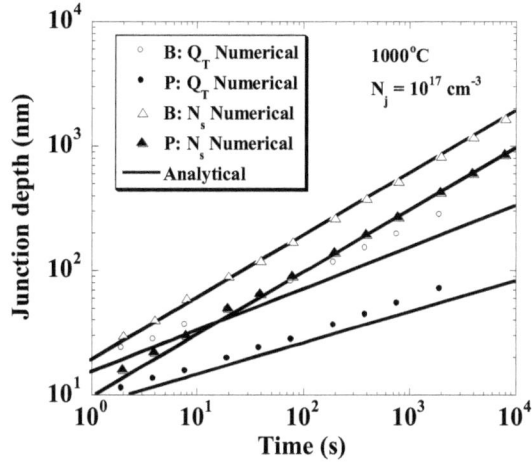

Figure 9: Dependence of junction depth on diffusion time. The constant total concentration and constant concentration diffusion source are shown.

Further, we assume in the analytical model that

$$D = \left(\frac{n}{n_i}\right)^{\gamma} D_i = \left(\frac{N}{n_i}\right)^{\gamma} D_i \tag{97}$$

However, this is not true in low contrition region, and is more rigorous one is

$$D = \left(\frac{n}{n_i}\right)^{\gamma} D_i = \begin{cases} \left(\dfrac{N}{n_i}\right)^{\gamma} D_i & for \ N > n_i \\ D_i & for \ N < n_i \end{cases} \tag{98}$$

Therefore, the diffusion coefficient in analytical model is smaller than that in numerical simulation that is

$$\left(\frac{N}{n_i}\right)^{\gamma} D_i < D_i \quad for \ N < n_i \tag{99}$$

This is the other reason why numerical junction depth is always deeper than the analytical one.

CONSTANT SURFACE CONCENTRATION

The analytical model for constant surface concentration was derived in [9].

We propose the form of s as

$$s = s_0\left(1 - a\xi - b\xi^2\right)^{\frac{1}{\gamma}} \tag{100}$$

Substituting Eq. 100 into Eq. 65, and inspecting s is independent of τ, we obtain

$$\frac{1}{\tau^{2k}}\frac{\partial}{\partial\xi}\left(s^\gamma\frac{\partial s}{\partial\xi}\right) + k\frac{\xi}{\tau}\frac{\partial s}{\partial\xi} = 0 \tag{101}$$

We impose that this is independent of τ, and obtain

$$k = \frac{1}{2} \tag{102}$$

Equation 101 is then reduced to

$$\frac{\partial}{\partial\xi}\left(s^\gamma\frac{\partial s}{\partial\xi}\right) + \frac{\xi}{2}\frac{\partial s}{\partial\xi} = 0 \tag{103}$$

Differentiating s with respect to ξ, we obtain

$$\frac{\partial s}{\partial\xi} = -\frac{s_0}{\gamma}\left(1 - a\xi - b\xi^2\right)^{\frac{1}{\gamma}-1}\left(a + 2b\xi\right) \tag{104}$$

Substituting Eq. 104 into Eq. 103, we obtain

$$\frac{\partial}{\partial\xi}\left[s_0{}^\gamma\left(1 - a\xi - b\xi^2\right)^{\frac{1}{\gamma}}\left(a + 2b\xi\right)\right] + \frac{\xi}{2}\left(1 - a\xi - b\xi^2\right)^{\frac{1}{\gamma}-1}\left(a + 2b\xi\right)$$

$$= s_0{}^\gamma\left[-\frac{1}{\gamma}\left(1 - a\xi - b\xi^2\right)^{\frac{1}{\gamma}-1}\left(a + 2b\xi\right)^2 + 2b\left(1 - a\xi - b\xi^2\right)^{\frac{1}{\gamma}}\right]$$

$$\quad + \frac{\xi}{2}\left(1 - a\xi - b\xi^2\right)^{\frac{1}{\gamma}-1}\left(a + 2b\xi\right)$$

$$= 0 \tag{105}$$

Dividing Eq. 105 by $\frac{1}{2}\left(1 - a\xi - b\xi^2\right)^{\frac{1}{\gamma}-1}$, we obtain

$$2s_0{}^\gamma\left[-\frac{1}{\gamma}(a+2b\xi)^2+2b(1-a\xi-b\xi^2)\right]+\xi(a+2b\xi)=0 \tag{106}$$

Ordering this with respect to power of ξ, we obtain

$$2s_0{}^\gamma(2\gamma b-a^2)+a[\gamma-4bs_0{}^\gamma(2+\gamma)]\xi+2b[\gamma-2bs_0{}^\gamma(2+\gamma)]\xi^2=0 \tag{107}$$

This should be valid for any ξ. Therefore, the followings should hold.

$$\xi^0:\ 2\gamma b=a^2 \tag{108}$$

$$\xi^1:\ b=\frac{\gamma}{4b(2+\gamma)s_0{}^\gamma} \tag{109}$$

$$\xi^2:\ b=\frac{\gamma}{2b(2+\gamma)s_0{}^\gamma} \tag{110}$$

However, Eqs. 109 and 110 contradict each other, and hence we cannot obtain a rigorous solution with this procedure. We hence try to find an approximate solution. We propose an approximate solution having a parameter η_γ as

$$2\gamma b=a^2 \tag{111}$$

$$b=\frac{\gamma}{\eta_\gamma 2(2+\gamma)s_0{}^\gamma} \tag{112}$$

Substituting Eqs. 111 and 112 into Eq. 107, we obtain

$$a\left[\gamma-4\frac{\gamma}{\eta_\gamma 2(2+\gamma)s_0{}^\gamma}s_0{}^\gamma(2+\gamma)\right]+2b\left[\gamma-2\frac{\gamma}{\eta_\gamma 2(2+\gamma)s_0{}^\gamma}s_0{}^\gamma(2+\gamma)\right]\xi$$

$$=a\left[\gamma-\frac{2\gamma}{\eta_\gamma}\right]+2b\left[\gamma-\frac{\gamma}{\eta_\gamma}\right]\xi$$

$$=a\left[\gamma-\frac{2\gamma}{\eta_\gamma}\right]+\frac{a^2}{\gamma}\left[\gamma-\frac{\gamma}{\eta_\gamma}\right]\xi=0$$

This is reduced to

$$1-\frac{2}{\eta_\gamma}+\frac{a}{\gamma}\left[1-\frac{1}{\eta_\gamma}\right]\xi=0 \tag{113}$$

We hope to evaluate average η_γ with respect to ξ. However, we evaluate the average with respect concentration. Defining a ratio as

$$\frac{s}{s_0} = r \tag{114}$$

we can relate it to ξ as

$$r^\gamma = 1 - a\xi - b\xi^2$$

Solving this with respect to ξ, we obtain

$$\xi = \frac{\gamma}{a}\left(\sqrt{1 + \frac{2}{\gamma}(1 - r^\gamma)} - 1 \right) \tag{115}$$

Substituting Eq. 115 into Eq. 113, we obtain

$$1 - \frac{2}{\eta_\gamma} + \left(\sqrt{1 + \frac{2}{\gamma}(1 - r^\gamma)} - 1 \right)\left[1 - \frac{1}{\eta_\gamma} \right] = 0$$

We then obtain

$$\eta_\gamma = 1 + \frac{1}{\sqrt{1 + \frac{2}{\gamma}(1 - r^\gamma)}} \tag{116}$$

Consequently, we obtain the average one as

$$\eta_{\gamma a} = 1 + \int_0^1 \frac{1}{\sqrt{1 + \frac{2}{\gamma}(1 - r^\gamma)}} dr \tag{117}$$

Let us consider the cases for γ of 1 and 2 as follows.

$$\begin{aligned}
\eta_{1a} &= 1 + \int_0^1 \frac{1}{\sqrt{1 + 2(1 - r)}} dr \\
&= 1 + \int_0^1 \frac{1}{\sqrt{3 - 2r}} dr \\
&= 1 - \frac{1}{2}\int_3^1 \frac{1}{\sqrt{l}} dl \qquad (for\ 3 - 2r = l) \\
&= \sqrt{3}
\end{aligned} \tag{118}$$

$$\eta_{2a} = 1 + \int_0^1 \frac{1}{\sqrt{1 + \left(1 - r^2\right)}} dr$$

$$= 1 + \left[\sin^{-1}\left(\frac{r}{\sqrt{2}}\right)\right]_0^1$$

$$= 1 + \frac{\pi}{4} \tag{119}$$

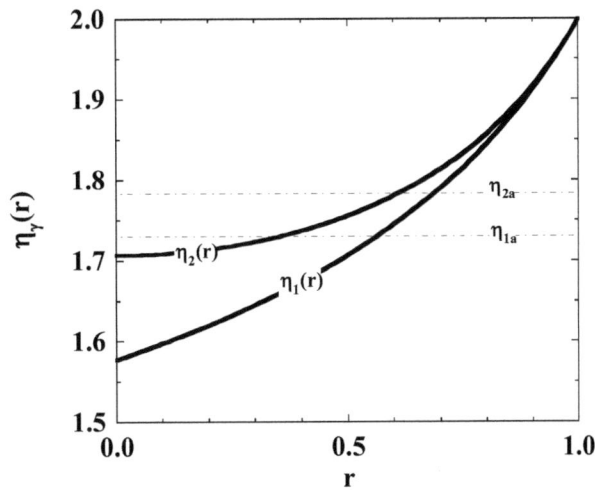

Figure 10: Dependence of $\eta_\gamma(r)$ and $\eta_{\gamma a}$ on r.

Fig. **10** shows the dependence of $\eta_\gamma(r)$ and $\eta_{\gamma a}$ on r. $\eta_\gamma(r)$ monotonically increases with increasing r, and $\eta_\gamma(r)$ varies between 1.5 and 2, and $\eta_{\gamma a}$ varies between 1.75 and 2. We can evaluate the limiting values as

$$\eta_\gamma(0) = 1 + \frac{1}{\sqrt{1 + \frac{2}{\gamma}}} \tag{120}$$

$$\eta_\gamma(1) = 2 \tag{121}$$

$\eta_\gamma(0)$ approaches to 2 with increasing γ while $\eta_\gamma(1)$ is invariable and is 2. Therefore, the analytical model can be expected to be accurate with increasing γ.

The diffusion profile is then expressed by

$$
\begin{aligned}
N &= N_s \left(1 - a\xi - b\xi^2 \right)^{\frac{1}{\gamma}} \\
&= N_s \left(1 - \sqrt{2\gamma b}\,\xi - b\xi^2 \right)^{\frac{1}{\gamma}} \\
&= N_s \left[1 - \sqrt{2\gamma}\,\sqrt{b}\,\xi - \left(\sqrt{b}\,\xi \right)^2 \right]^{\frac{1}{\gamma}} \\
&= N_s \left[1 - \sqrt{2\gamma}\,Y_\gamma - Y_\gamma^2 \right]^{\frac{1}{\gamma}}
\end{aligned}
\tag{122}
$$

where normalized distance Y_γ is defined by

$$
\begin{aligned}
Y_\gamma &= \sqrt{ \frac{\gamma}{2(2+\gamma)\eta_{\gamma a} s_0{}^\gamma} } \, \frac{x}{\sqrt{2 D_i t}} \\
&= \frac{x}{\sqrt{ \dfrac{2+\gamma}{\gamma} \eta_{\gamma a} \left(\dfrac{N_s}{n_i} \right)^\gamma 4 D_i t }}
\end{aligned}
\tag{123}
$$

We can evaluate the junction depth by the depth where $N = 0$ given by

$$
1 - \sqrt{2\gamma}\,Y_\gamma - Y_\gamma^2 = 0
\tag{124}
$$

We obtain related normalized distance as

$$
Y_\gamma = \frac{\sqrt{2\gamma + 4} - \sqrt{2\gamma}}{2} = \frac{x_j}{\sqrt{ \dfrac{2(2+\gamma)\eta_{\gamma a}}{\gamma} \left(\dfrac{N_s}{n_i} \right)^\gamma 2 D_i t }}
\tag{125}
$$

and corresponding junction depth is given by

$$
x_j = \frac{\sqrt{2\gamma + 4} - \sqrt{2\gamma}}{2} \sqrt{ \frac{2+\gamma}{\gamma} \eta_{\gamma a} \left(\frac{N_s}{n_i} \right)^\gamma 4 D_i t }
\tag{126}
$$

It should be noted that it is proportional to $\sqrt{t}$ independent of γ.

In the simulation, we treat B for $\gamma = 1$, and phosphorous for $\gamma = 2$.

Fig. **11** compares analytical and numerical B diffusion profiles with $\gamma = 1$. The analytical model agrees well with numerical one. The diffusing profile in Y_1 plot is time invariant.

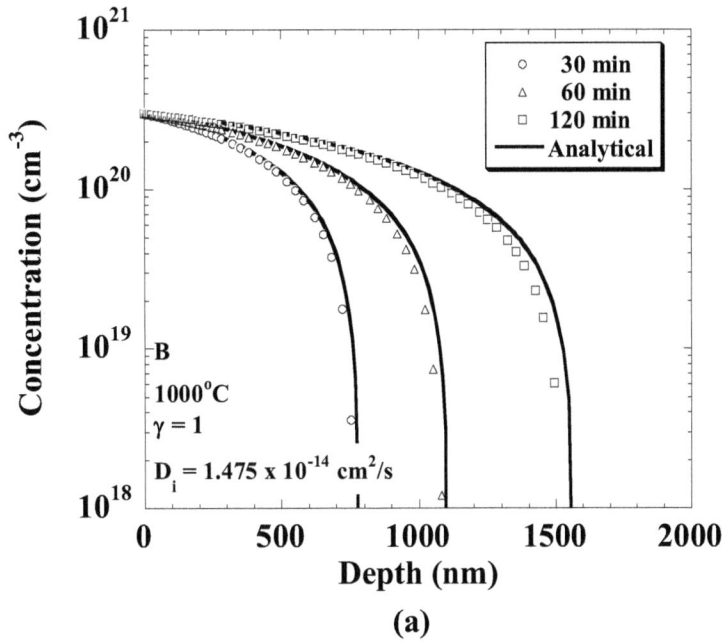

(a)

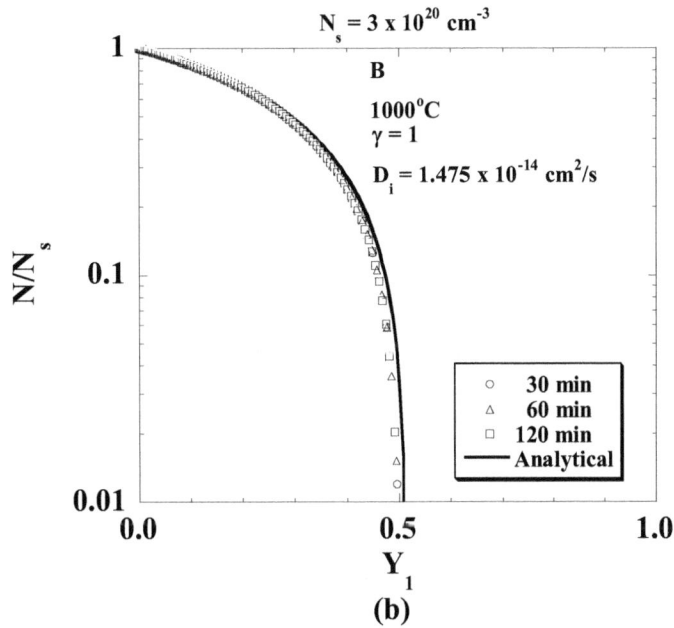

(b)

Figure 11: B diffusion profile ($\gamma = 1$) (a) Real distance (b) Normalized distance.

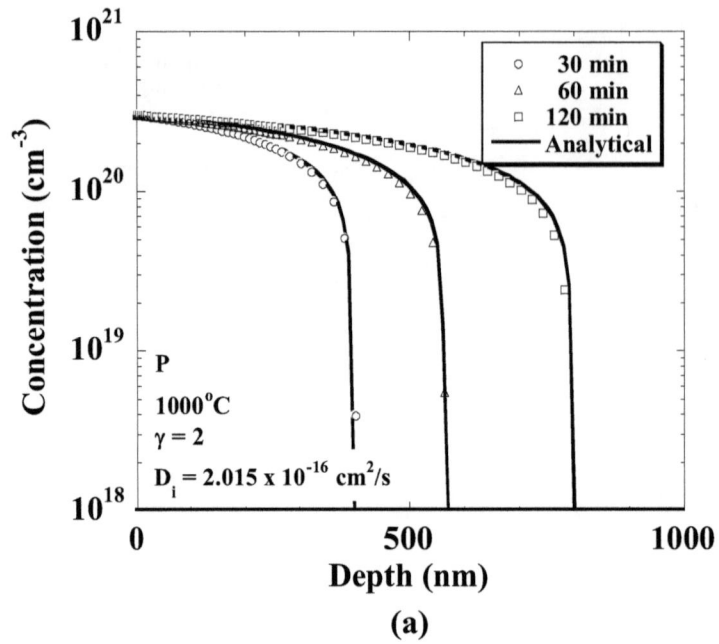

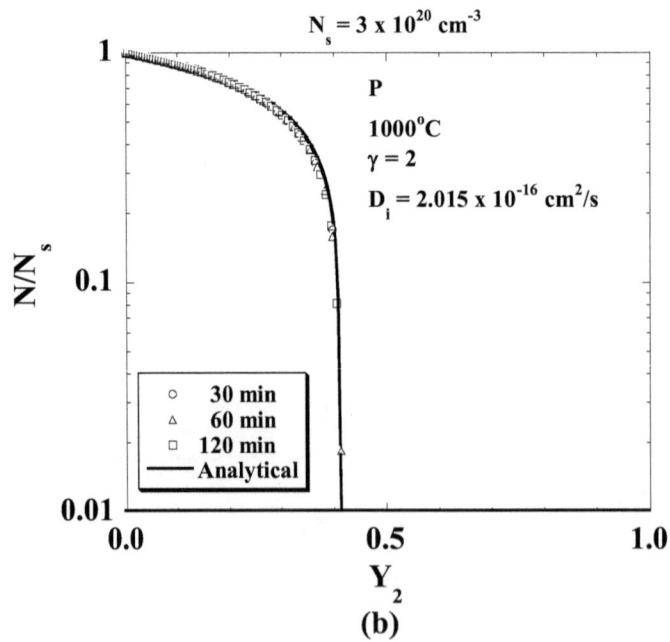

Figure 12: Dependence of P diffusion profile ($\gamma = 2$) (a) Real distance. (b) Normalized distance.

Fig. **12** compares P diffusion profiles with $\gamma = 2$. The analytical model agrees well with numerical one. The diffusion profiles become the same in Y_2 plot. The profile looks more box-like than that of B.

Fig. **13** shows dependence of diffusion profiles on diffusion time in normalized plot. It is clear that the profile becomes box like with increasing γ.

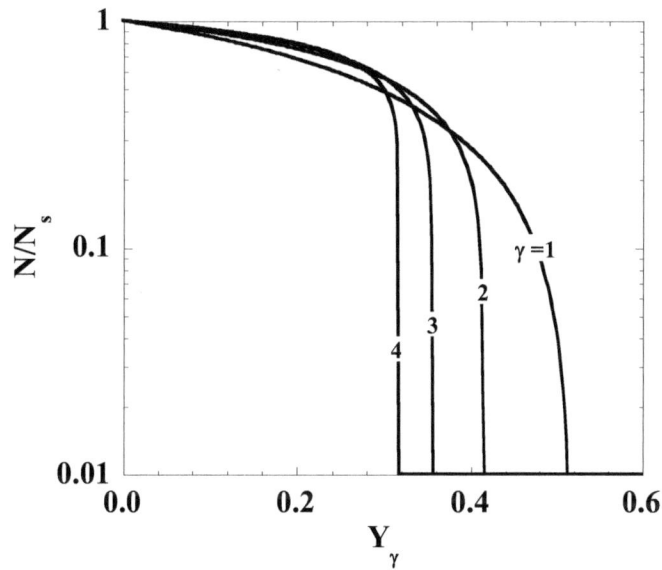

Figure 13: Dependence of normalized diffusion profile on γ.

COMPARISON WITH EXPERIMENTAL DATA

We compare analytical model and experimental data here.

Sub-keV ion implantation is frequently used in modern VLSI processes to realize shallow junctions. We compared the experimental data of these samples with the analytical model [9].

In these cases, constant dose case seems to be applicable. However, solid solubility limits the maximum active concentration, and it plays a role of constant concentration diffusion source. Therefore, we apply a model for constant dose.

We should also consider transient enhanced diffusion (TED) in the early stage of diffusion. However, our model does not include TED. We propose to introduce a parameter t_l to effectively express TED.

Fig. **14** compares the analytical experimental diffusion profile. The model readily expresses the experimental data with one parameter set. We use $\gamma = 0.3$ for B and P. According to the diffusion mechanism, γ should be a natural number, and 0.3 of γ means some mechanisms contribute to the diffusion. We use $\gamma = 1$ for As, and single negative paring diffusion mechanism dominates the diffusion.

TWO-DIMENSIONAL DIFFUSION PROFILES

We investigate two-dimensional diffusion profiles in gate patterned substrate. It is important to know the relationship between vertical and lateral diffusion distribution. In this section, we only consider the case where the diffusion coefficients are constant.

DIFFUSION PROFILE WITH POINT DIFFUSION SOURCE

We first derive diffusion distribution function with point diffusion source in infinite plane.

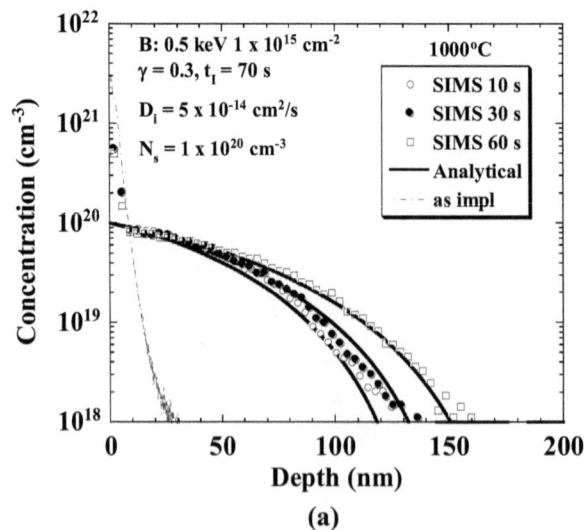

(a)

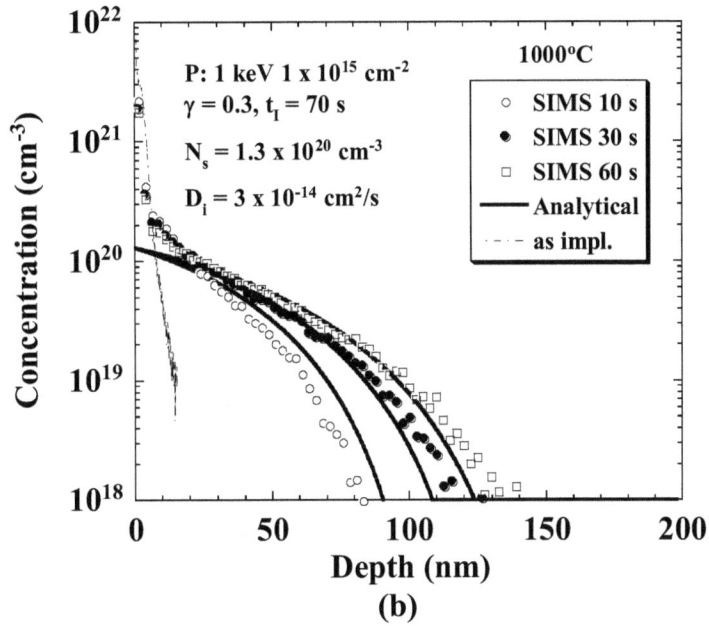

(b)

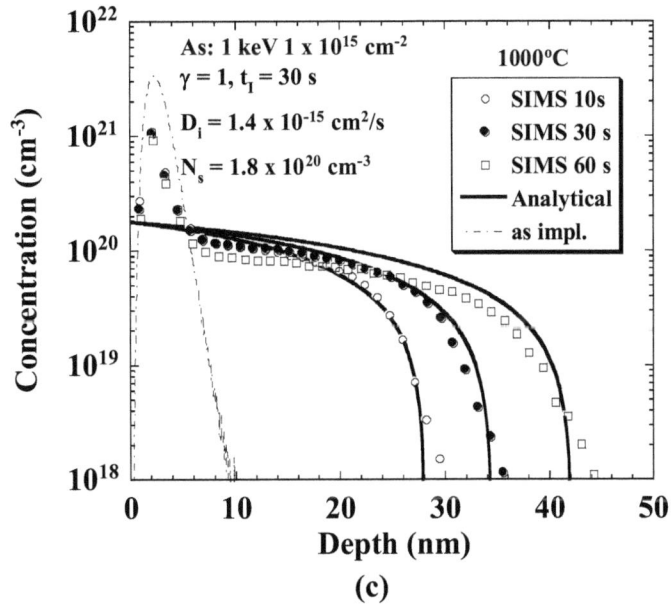

(c)

Figure 14: Comparison between analytical and experimental diffusion profiles. (a) B, (b) P, (c) As

The diffusion equation in spherical system is given by

$$\frac{\partial N}{\partial t} = D\left(\frac{d^2 N}{dr^2} + \frac{2}{r}\frac{\partial N}{\partial r}\right) \tag{127}$$

We neglect angular dependence in Eq. 127. We assume spherical initial condition with radius of a instead of point diffusion source given by

$$N(0, r) = \begin{cases} N_0 & \text{for } 0 \le r < a \\ 0 & \text{for } r > a \end{cases} \tag{128}$$

The total dose Φ is given by

$$\Phi = \frac{4}{3}\pi a^3 N_0 \tag{129}$$

We introducing a variable

$$U = rN \tag{130}$$

The differentiation of this variable is as follows.

$$\frac{\partial N}{\partial r} = \frac{rU' - U}{r^2} \tag{131}$$

$$\frac{d^2 N}{dr^2} = \frac{r^3 U'' - 2r^2 U' + 2rU}{r^4} \tag{132}$$

Substituting Eq. 131 and 132 into Eq. 127, we obtain a differential equation with respect to U as

$$\frac{\partial U}{\partial t} = D\frac{d^2 U}{dr^2} \tag{133}$$

The corresponding initial condition is

$$U(0, r) = \begin{cases} rN_0 & \text{for } 0 \le r < a \\ 0 & \text{for } r > a \end{cases} \tag{134}$$

The boundary condition is

$$U(t, 0) = 0 \tag{135}$$

Performing Fourier transformation to Eq. 133 with respect to r, we obtain

$$\frac{d}{dt} \mathscr{F}\{U(t,r)\} = -D\xi^2 \mathscr{F}\{U\} \tag{136}$$

r is valid only when it is positive, but we tentatively assume it is valid in infinite region. Equation 136 is then solved as

$$\mathscr{F}\{U(t,r)\} = A(\xi)\exp(-D\xi^2 t) \tag{137}$$

Performing Fourier transformation to the initial condition, we obtain

$$\mathscr{F}\{U(0,r)\} = A(\xi) \tag{138}$$

Substituting Eq. 138 into Eq. 137, we obtain

$$\begin{aligned}
\mathscr{F}\{U(t,r)\} &= \mathscr{F}\{U(0,r)\}\exp(-D\xi^2 t) \\
&= \mathscr{F}\{U(0,r)\}\mathscr{F}\left\{\frac{1}{\sqrt{2Dt}}\exp\left(-\frac{r^2}{4Dt}\right)\right\}
\end{aligned} \tag{139}$$

Performing reverse Fourier transformation, we obtain

$$U(t,r) = \frac{1}{\sqrt{2\pi}}\int_{-\infty}^{\infty} U(0,\xi)\,\frac{1}{\sqrt{2Dt}}\exp\left[-\frac{(r-\xi)^2}{4Dt}\right]d\xi \tag{140}$$

We set $U(0,\xi) = 0$ when ξ is negative. This solution satisfies the diffusion equation and initial condition, but does not satisfy boundary condition. We virtually add initial condition in the negative ξ region as

$$U(0,r) = \begin{cases} rN_0 & \text{for } -a \leq r < 0 \\ 0 & \text{for } r < -a \end{cases} \tag{141}$$

That is, we add a function that satisfies $U(0, -r) = -U(0, r)$, which enable us to satisfy the boundary condition. Equation 140 is then modified as

$$U(t, r) = \frac{1}{2\sqrt{\pi Dt}}\left[\int_{-\infty}^{0} U(0, \xi)\exp\left[-\frac{(r-\xi)^2}{4Dt}\right]d\xi + \int_{0}^{\infty} U(0, \xi)\exp\left[-\frac{(r-\xi)^2}{4Dt}\right]d\xi\right]$$

$$= \frac{1}{2\sqrt{\pi Dt}}\left[-\int_{\infty}^{0} U(0, -s)\exp\left[-\frac{(r+s)^2}{4Dt}\right]ds + \int_{0}^{\infty} U(0, \xi)\exp\left[-\frac{(r-\xi)^2}{4Dt}\right]d\xi\right]$$

$$= \frac{1}{2\sqrt{\pi Dt}}\left[\int_{\infty}^{0} U(0, s)\exp\left[-\frac{(r+s)^2}{4Dt}\right]ds + \int_{0}^{\infty} U(0, \xi)\exp\left[-\frac{(r-\xi)^2}{4Dt}\right]d\xi\right]$$

$$= \frac{1}{2\sqrt{\pi Dt}}\left[\int_{0}^{\infty} U(0, \xi)\left\{\exp\left[-\frac{(r-\xi)^2}{4Dt}\right] - \exp\left[-\frac{(r+\xi)^2}{4Dt}\right]\right\}d\xi\right]$$

$$= \frac{1}{2\sqrt{\pi Dt}}\left[\int_{0}^{a} N_0\xi\left\{\exp\left[-\frac{(r-\xi)^2}{4Dt}\right] - \exp\left[-\frac{(r+\xi)^2}{4Dt}\right]\right\}d\xi\right] \tag{142}$$

Keeping Φ constant with varying N_0, we make $a \to 0$, and hence we expand exponential term in 142 into Taylor series and select only first term, and obtain

$$U(t, r) = \frac{N_0}{2\sqrt{\pi Dt}}\exp\left[-\frac{r^2}{4Dt}\right]\left[\int_{0}^{a} \xi\left\{\left[1 + \frac{2r}{4Dt}\xi\right] - \left[1 - \frac{2r}{4Dt}\xi\right]\right\}d\xi\right]$$

$$= \frac{N_0}{2\sqrt{\pi Dt}}\exp\left[-\frac{r^2}{4Dt}\right]\left[\int_{0}^{a} \frac{r}{Dt}\xi^2 d\xi\right]$$

$$= \frac{N_0 r a^3}{6\sqrt{\pi Dt}Dt}\exp\left[-\frac{r^2}{4Dt}\right] \tag{143}$$

Using a dose Φ, we obtain

$$U(t, r) = \frac{r\Phi}{(2\sqrt{\pi Dt})^3}\exp\left[-\frac{r^2}{4Dt}\right] \tag{144}$$

Consequently, we obtain a diffusion profile with point diffusion source located at r_0 as

$$N = \frac{\Phi}{(2\sqrt{\pi Dt})^3}\exp\left[-\frac{(r-r_0)^2}{4Dt}\right] \tag{145}$$

TWO-DIMENSIONAL DIFFUSION PROFILES IN MOS STRUCTURE SUBSTRATE

We first consider the impurities in the drain region. We set the origin of lateral axis at the gate edge.

We regards that the point diffusion source exist on the drain region. That is, we set them at $(x_i, 0, z_i)$ where $0 < x_i < \infty$ and $-\infty < z_i < \infty$.

The contribution of the increment square area $(x_i, 0, z_i), (x_i + dx_i, 0, z_i + dz_i)$ to the concentration at (x, y, z) is given by

$$dN(r, t) = \frac{\Phi}{\left(2\sqrt{\pi Dt}\right)^3} \exp\left[-\frac{\left(x - x_i\right)^2 + y^2 + \left(z - z_i\right)^2}{4Dt}\right] dx_i dz_i \tag{146}$$

Integrating all contribution, we obtain

$$N(r, t) = \frac{\Phi}{\left(2\sqrt{\pi Dt}\right)^3} \exp\left[-\frac{y^2}{4Dt}\right] \int_0^\infty \exp\left[-\frac{\left(x - x_i\right)^2}{4Dt}\right] dx_i \int_{-\infty}^\infty \exp\left[-\frac{\left(z - z_i\right)^2}{4Dt}\right] dz_i$$

$$= \frac{\Phi}{2\sqrt{\pi Dt}} \exp\left[-\frac{y^2}{4Dt}\right] \frac{1 + \mathrm{erf}\left(\frac{x}{2\sqrt{Dt}}\right)}{2} \tag{147}$$

Note that Eq. 147 does not depend on z, this is due to that we integrate in the region $[-\infty, \infty]$.

In the limiting case of $x \to \infty$, we obtain one-dimensional profile given by

$$N(y) = \frac{\Phi}{2\sqrt{\pi Dt}} \exp\left[-\frac{y^2}{4Dt}\right] \tag{148}$$

as expected.

Equation 147 expresses two-dimensional diffusion profiles assuming point diffusion source on the drain region. The model should be improved as the following.

We locate the source at the surface, but we usually use ion implantation, and the peak position is generally located in the substrate.

Further, we assume infinite plane, and the impurities are assumed to diffuse in the air region as if the air is the substrate.

We modify Eq. 147 to include the above discussion.

First, we set the diffusion source at the locating at R_p, and the model is then

$$N(x, y) = \frac{\Phi}{2\sqrt{\pi D t}} \exp\left[-\frac{(y - R_p)^2}{4Dt}\right] \frac{1 + \mathrm{erf}\left(\frac{x}{2\sqrt{Dt}}\right)}{2} \tag{149}$$

In the limiting case of $x \to \infty$, the diffusion profile becomes on dimensional one given by

$$N(y) = \frac{\Phi}{2\sqrt{\pi D t}} \exp\left[-\frac{(y - R_p)^2}{4Dt}\right] \tag{150}$$

We usually introduce impurities by ion implantation, which is approximately expressed with Gaussian profile given by

$$N(y) = \frac{\Phi}{\sqrt{2\pi \Delta R_p^2}} \exp\left[-\frac{(y - R_p)^2}{2\Delta R_p^2}\right] \tag{151}$$

We regard Eq. 150 becomes Eq. 151 after the certain time period t_0 as

$$\frac{\Phi}{\sqrt{2\pi \Delta R_p^2}} \exp\left[-\frac{(y - R_p)^2}{2\Delta R_p^2}\right] = \frac{\Phi}{\sqrt{2\pi 2Dt_0}} \exp\left[-\frac{(y - R_p)^2}{4Dt}\right] \tag{152}$$

Therefore, we obtain

$$t_0 = \frac{\Delta R_p^2}{2D} \tag{153}$$

On the other hand, the function related to the lateral distribution is error function for both ion implantation and diffusion. We set the related offset time as t_{0t}, and obtain

$$2\sqrt{Dt_{0t}} = \sqrt{2}\Delta R_{pt}$$

Therefore, the offset time associated with lateral distribution is given by

$$t_{0t} = \frac{\Delta R_{pt}^{2}}{2D} \tag{154}$$

We then obtain the diffusion profile as

$$N(x, y) = \frac{\Phi}{2\sqrt{\pi D(t + t_0)}} \exp\left[-\frac{(y - R_p)^2}{4D(t + t_0)}\right] \frac{1 + \mathrm{erf}\left(\dfrac{x}{2\sqrt{D(t + t_{0t})}}\right)}{2}$$

$$= \frac{\Phi}{\sqrt{2\pi(2Dt + \Delta R_p^{2})}} \exp\left[-\frac{(y - R_p)^2}{2(2Dt + \Delta R_p^{2})}\right] \frac{1 + \mathrm{erf}\left(\dfrac{x}{\sqrt{2(2Dt + \Delta R_{pt}^{2})}}\right)}{2} \tag{155}$$

Further, we want to make the flux at the surface zero. This can be realized by that we set the same diffusion source at $y = -R_p$. We then obtain

$$N(x, y) = \frac{\Phi}{\sqrt{2\pi(2Dt + \Delta R_p^{2})}} \left\{\exp\left[-\frac{(y - R_p)^2}{2(2Dt + \Delta R_p^{2})}\right] + \exp\left[-\frac{(y + R_p)^2}{2(2Dt + \Delta R_p^{2})}\right]\right\} \times$$

$$\frac{1 + \mathrm{erf}\left(\dfrac{x}{\sqrt{2(2Dt + \Delta R_{pt}^{2})}}\right)}{2} \tag{156}$$

This is the diffusion distribution in the drain region. We move the origin from the gate edge to the center of the gate of length L_G and obtain

$$N_D(x, y) = \frac{\Phi}{\sqrt{2\pi(2Dt + \Delta R_p^{2})}} \left\{\exp\left[-\frac{(y - R_p)^2}{2(2Dt + \Delta R_p^{2})}\right] + \exp\left[-\frac{(y + R_p)^2}{2(2Dt + \Delta R_p^{2})}\right]\right\} \times$$

$$\frac{1 + \mathrm{erf}\left(\dfrac{x - \dfrac{L_G}{2}}{\sqrt{2(2Dt + \Delta R_{pt}^{2})}}\right)}{2} \tag{157}$$

Similar analysis can be applied to the impurities in source region, and we can obtain

$$N_S(x, y) = \frac{\Phi}{\sqrt{2\pi\left(2Dt + \Delta R_p^{\;2}\right)}} \left\{ \exp\left[-\frac{\left(y - R_p\right)^2}{2\left(2Dt + \Delta R_p^{\;2}\right)}\right] + \exp\left[-\frac{\left(y + R_p\right)^2}{2\left(2Dt + \Delta R_p^{\;2}\right)}\right] \right\} \times$$

$$\frac{1 + \mathrm{erf}\left(-\dfrac{x + \dfrac{L_G}{2}}{\sqrt{2\left(2Dt + \Delta R_{pt}^{\;2}\right)}}\right)}{2}$$

$$\tag{158}$$

Consequently, we obtain an analytical model for total impurity diffusion profile as

$$N_{S/D}(x, y) = N_S(x, y) + N_D(x, y)$$

$$= \frac{\Phi}{\sqrt{2\pi\left(2Dt + \Delta R_p^{\;2}\right)}} \left\{ \exp\left[-\frac{\left(y - R_p\right)^2}{2\left(2Dt + \Delta R_p^{\;2}\right)}\right] + \exp\left[-\frac{\left(y + R_p\right)^2}{2\left(2Dt + \Delta R_p^{\;2}\right)}\right] \right\} \times$$

$$\left[1 - \frac{\mathrm{erf}\left(\dfrac{\dfrac{L_G}{2} - x}{\sqrt{2\left(2Dt + \Delta R_{pt}^{\;2}\right)}}\right) + \mathrm{erf}\left(\dfrac{\dfrac{L_G}{2} + x}{\sqrt{2\left(2Dt + \Delta R_{pt}^{\;2}\right)}}\right)}{2} \right]$$

$$\tag{159}$$

If we want to obtain profiles with zero surface boundary condition, we set the negative value diffusion source at $y = -R_p$ and obtain

$$N_{S/D}(x, y) = \frac{\Phi}{\sqrt{2\pi\left(2Dt + \Delta R_p^{\;2}\right)}} \left\{ \exp\left[-\frac{\left(y - R_p\right)^2}{2\left(2Dt + \Delta R_p^{\;2}\right)}\right] - \exp\left[-\frac{\left(y + R_p\right)^2}{2\left(2Dt + \Delta R_p^{\;2}\right)}\right] \right\} \times$$

$$\left[1 - \frac{\mathrm{erf}\left(\dfrac{\dfrac{L_G}{2} - x}{\sqrt{2\left(2Dt + \Delta R_{pt}^{\;2}\right)}}\right) + \mathrm{erf}\left(\dfrac{\dfrac{L_G}{2} + x}{\sqrt{2\left(2Dt + \Delta R_{pt}^{\;2}\right)}}\right)}{2} \right]$$

$$\tag{160}$$

We can apply this procedure to any patterned substrate. Let us consider contact pattern. We can easily obtain the diffusion profiles in an infinite plane as

$$N(r, t) = \frac{\Phi}{\left(2\sqrt{\pi Dt}\right)^3} \exp\left[-\frac{y^2}{4Dt}\right] \int_{-\frac{L}{2}}^{\frac{L}{2}} \exp\left[-\frac{\left(x - x_i\right)^2}{4Dt}\right] dx_i \int_{-\infty}^{\infty} \exp\left[-\frac{\left(z - z_i\right)^2}{4Dt}\right] dz_i$$

$$= \frac{\Phi}{2\sqrt{\pi Dt}} \exp\left[-\frac{y^2}{4Dt}\right] \frac{\mathrm{erf}\left(\dfrac{\dfrac{L}{2} - x}{2\sqrt{Dt}}\right) + \mathrm{erf}\left(\dfrac{\dfrac{L}{2} + x}{2\sqrt{Dt}}\right)}{2}$$

$$\tag{161}$$

The solution with zero flux at the surface is given by

$$N(x, y) = \frac{\Phi}{\sqrt{2\pi\left(2Dt + \Delta R_p^2\right)}} \left\{ \exp\left[-\frac{\left(y - R_p\right)^2}{2\left(2Dt + \Delta R_p^2\right)}\right] + \exp\left[-\frac{\left(y + R_p\right)^2}{2\left(2Dt + \Delta R_p^2\right)}\right] \right\} \times$$

$$\frac{\operatorname{erf}\left(\dfrac{\frac{L}{2} - x}{\sqrt{2\left(2Dt + \Delta R_{pt}^2\right)}}\right) + \operatorname{erf}\left(\dfrac{\frac{L}{2} + x}{\sqrt{2\left(2Dt + \Delta R_{pt}^2\right)}}\right)}{2}$$

$$(162)$$

The solution with infinite flux at the surface is given by

$$N(x, y) = \frac{\Phi}{\sqrt{2\pi\left(2Dt + \Delta R_p^2\right)}} \left\{ \exp\left[-\frac{\left(y - R_p\right)^2}{2\left(2Dt + \Delta R_p^2\right)}\right] - \exp\left[-\frac{\left(y + R_p\right)^2}{2\left(2Dt + \Delta R_p^2\right)}\right] \right\} \times$$

$$\frac{\operatorname{erf}\left(\dfrac{\frac{L}{2} - x}{\sqrt{2\left(2Dt + \Delta R_{pt}^2\right)}}\right) + \operatorname{erf}\left(\dfrac{\frac{L}{2} + x}{\sqrt{2\left(2Dt + \Delta R_{pt}^2\right)}}\right)}{2}$$

$$(163)$$

Fig. **15** shows the two-dimensional diffusion profiles of B ion implanted at 10 keV and a dose of $1\times10^{15}\,cm^{-2}$ with gate length of 0.5 μm. 1000°C for 5 minutes annealing was performed. The corresponding two-dimensional profiles are evaluated with the model implemented in FabMeister-IM [13]. We can obtain simple physical intuition with this model.

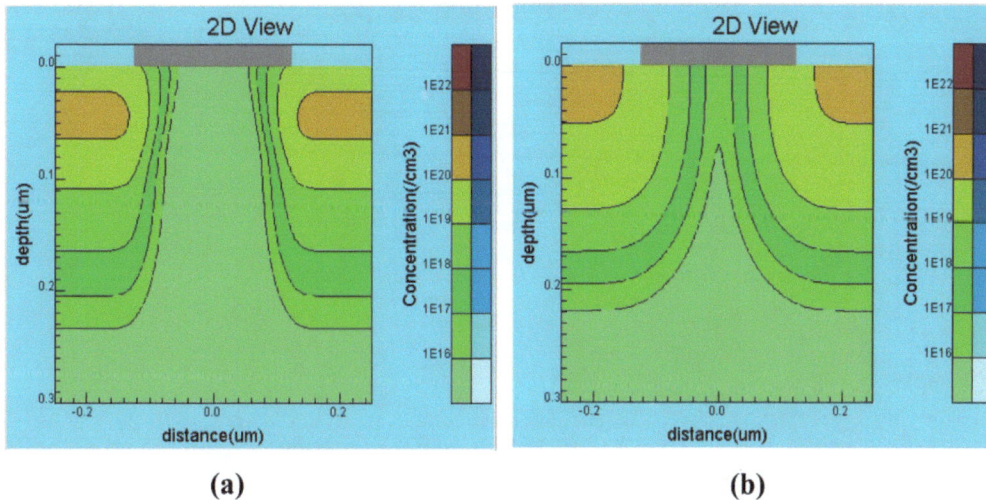

Figure 15: Two-dimensional impurity diffusion profiles with gate pattern substrate. (a) as implanted. (b) 1000°C for 5 min annealed.

DIFFUSION IN MULCH LAYER

p[+]polycrystalline silicon (polysilicon) gate is used for pMOSFETs and the gate is fabricated by B or BF_2 ion implantation into the polysilicon gate and subsequent annealing. We should be careful about the annealing condition with two aspects described below.

Deep submicron MOSFETs require thin gate oxide. B possibly penetrates this thin oxide and induces the threshold voltage variation. Therefore, we should set the annealing condition to alleviate the penetration.

The ion implantation profile in polysilicon is non-uniform and the concentration at the polysilicon/gate SiO_2 interface is low. This concentration should be increased to suppress gate depletion, that is, we want to obtain flat impurity concentration in gate polysilicon.

Consequently, we should set the process condition so that the profile in polysilicon is uniform without B penetration through the gate oxide [14].

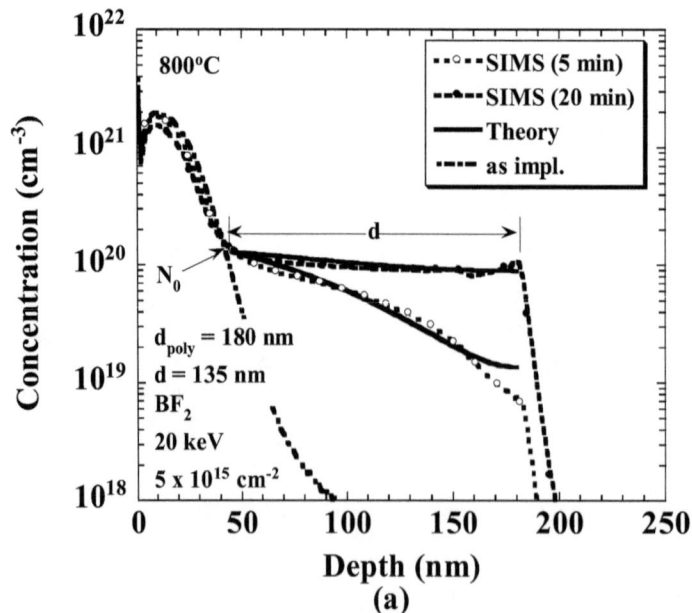

(a)

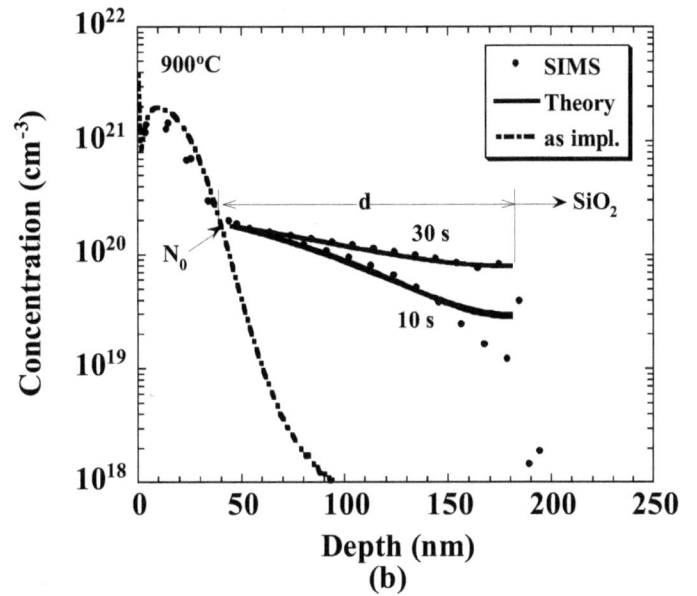

(b)

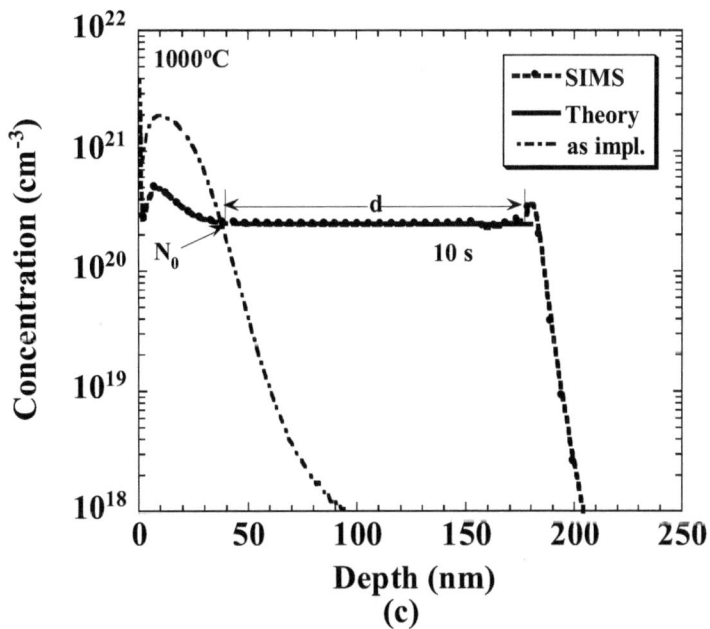

(c)

Figure 16: B diffusion profile in polysilicon gate. (a) 800°C. (b) 900°C (C) 1000°C.

DIFFUSION IN POLYSILICON GATE

First we consider diffusion in polysilicon gate as shown in Fig. **16**, where 180 nm thick gate polysilicon formed and BF$_2$ are ion implanted into it at 20 keV and a dose of 5×10^{15} cm^{-2}. B is immobile at high concentration region in the high concentration region of more 10^{20} cm^{-3}. The profiles of each temperature can be regarded as those with a boundary condition of constant surface concentration N_0, and its temperature dependence is expressed by

$$N_0 = 8.19 \times 10^{21} \exp\left[-\frac{0.38(eV)}{k_B T} \right] cm^{-3} \tag{164}$$

The diffusion equation in this polysilicon gate can be described as

$$\frac{\partial N(x,t)}{\partial t} = D_{poly} \frac{\partial^2 N(x,t)}{\partial^2 x} \tag{165}$$

where constant diffusion coefficient in polysilicon gate D_{poly} is assumed. The corresponding boundary conditions are given by

$$\begin{cases} N(0,t) = N_0 \\ \left. \dfrac{\partial N(x,t)}{\partial x} \right|_{x=d} = 0 \end{cases} \tag{166}$$

The origin is not set at the surface, but the kink point for the diffusion as defined in Fig. **16**, where as-implanted concentration is N_0. d is the distance from the origin to the polysilicon/gate oxide interface. We also assume the zero flux at the polysilicon/gate oxide interface. We also assume zero initial condition given by

$$N(x,0) = 0 \tag{167}$$

which is not true. The approximation means that the as-implanted concentration in tail region deeper than the origin is set zero. D_{poly} is high and the tail region is covered by the diffused impurities from the diffusion source in quite short time, and hence the approximation is always good.

The above diffusion equations obviously have a solution in the infinite time as

$$N(x,\infty)=N_0 \tag{168}$$

We therefore introduce a variable of

$$n(x,t)=N_0-N(x,t) \tag{169}$$

The differential equations for $n(x,t)$ is then given by

$$\frac{\partial n(x,t)}{\partial t}=D_{poly}\frac{\partial^2 n(x,t)}{\partial^2 x} \tag{170}$$

$$\begin{cases} n(0,t)=0 \\ \left.\dfrac{\partial n(x,t)}{\partial x}\right|_{x=d}=0 \end{cases} \tag{171}$$

$$n(x,0)=N_0 \tag{172}$$

We assume that the solution have a form of product of independent function of $X(x)$ and $T(t)$ given by

$$n(x,t)=X(x)T(t) \tag{173}$$

Substituting Eq. 173 into Eq. 170, and divide both side by $D_{poly}XT$, we obtain

$$\frac{T'}{D_{poly}T}=\frac{X''}{X} \tag{174}$$

This form imposes that both sides should depend on neither t nor x. We hence set them as $-\lambda^2\,(\lambda>0)$. We should assume the form for constant number to alleviate infinite solution for infinite time. Equation 174 is then reduced to

$$\begin{cases} T'+\lambda^2 D_{poly}T=0 \\ X''+\lambda^2 X=0 \end{cases} \tag{175}$$

We then obtain a solution of

$$n(x,t) = \exp\left(-D_{poly}\lambda^2 t\right)\left[A\sin\left(\lambda x\right) + B\cos\left(\lambda x\right)\right] \tag{176}$$

where A and B are arbitrary constants.

Substituting Eq. 176 into the first one of Eq. 171, we obtain

$$n(0,t) = B\exp\left(-D_{poly}\lambda^2 t\right) = 0 \tag{177}$$

This leads to $B = 0$. We further substitute Eq. 176 with $B = 0$ into the second one of eq. 171, and obtain

$$A\sin\left(\lambda d\right)\exp\left(-D_{poly}\lambda^2 t\right) = 0 \tag{178}$$

Therefore, we have many values for λ that hold Eq. 178 as

$$\lambda_k = \frac{(1+2k)\pi}{2d} \quad for\ k = 0,1,2,\cdots \tag{179}$$

Therefore, the solution has a form of

$$n(x,t) = \sum_{k=0}^{\infty} A_k \exp\left(-D_{poly}\lambda_k^2 t\right)\sin\left(\lambda_k x\right) \tag{180}$$

where A_k is the arbitrary constant related to k. Substituting Eq. 180 into the initial condition of Eq. 172, we obtain

$$n(x,0) = N_0 = \sum_{k=0}^{\infty} A_k \sin\left(\lambda_k x\right) \tag{181}$$

Multiplying $\sin\left(\lambda_l x\right)$ to both sides of Eq. 181, and integrating it with respect to x from 0 and d, we obtain

$$\begin{aligned}
N_0 \int_0^d \sin\left(\lambda_l x\right)dx &= \sum_{k=0}^{\infty} A_k \int_0^d \sin\left(\lambda_k x\right)\sin\left(\lambda_l x\right)dx \\
&= A_l \int_0^d \sin^2\left(\lambda_l x\right)dx
\end{aligned} \tag{182}$$

We utilize that right side integration is not zero only when $k = l$. Performing the integration, we obtain

$$A_l = \frac{4N_0}{(1+2l)\pi} \tag{183}$$

Therefore, we obtain the time dependent profile in polysilicon gate as

$$N(x,t) = N_0 - \sum_{k=0}^{\infty} A_l \frac{4N_0}{(1+2k)\pi} \exp\left(-\frac{t}{\tau_k}\right) \sin(\lambda_k x) \tag{184}$$

where

$$\tau_k = \frac{1}{\lambda_k^2 D_{poly}}$$

$$= \left[\frac{2d}{(1+2k)\pi}\right]^2 \frac{1}{D_{poly}} \tag{185}$$

The theory agrees well with experimental data as shown in Fig. **16**, where the diffusion coefficient in gate polysilicon is extracted as

$$D_{poly} = 2.33 \times 10^3 \exp\left[-\frac{3.49(eV)}{k_B T}\right] cm^2\!\!\Big/\!s \tag{186}$$

When the diffusion proceeds further and the profile becomes almost flat, only the first term dominates profile given by

$$N(x,t) = N_0 - \frac{4N_0}{\pi} \exp\left(-\frac{t}{\tau_0}\right) \sin(\lambda_0 x) \tag{187}$$

We can evaluate the flatness of the profile as $N(d)\!\!\Big/\!_{N_0} = r$. Using Eq. 187, we can evaluate the critical time t_1 for flattening the profile as

$$t_1 = \tau_0 \ln\left[\frac{4}{(1-r)\pi}\right] \tag{188}$$

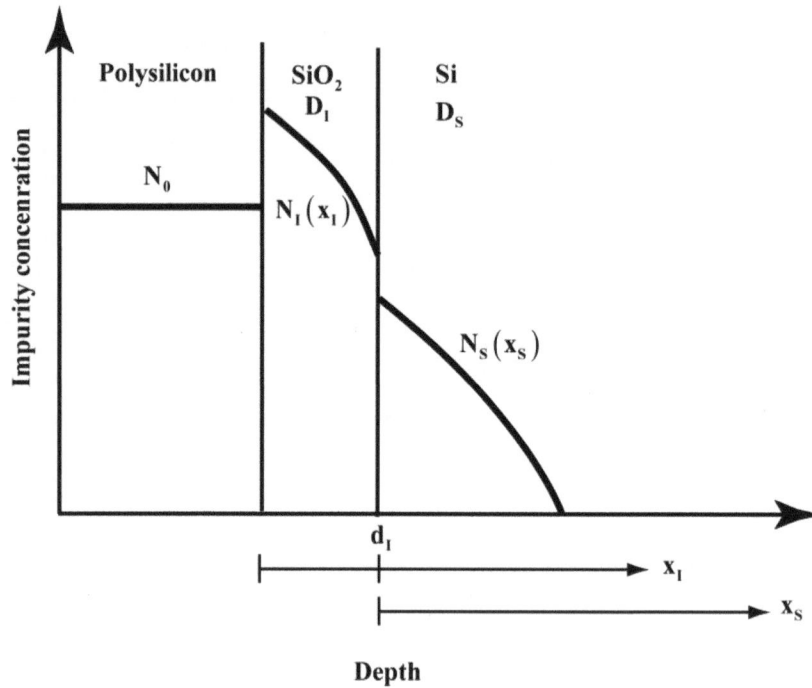

Figure 17: An axis system for impurity diffusion profiles in polysilicon gate/gate oxide/Si substrate structure.

DIFFUSION THROUGH GATE OXIDE

Next we consider diffusion through gate SiO_2. The analytical axis for the system is shown in Fig. **17**, where d_I is the gate oxide thickness. We assign axis related to gate oxide region as x_I, and one related to Si substrate region as x_S, and related origin are set as shown in the figure.

The concentration in gate polysilicon N_G is assumed to be constant and is N_0, that is

$$N_G\left(x,t\right) = N_0 \tag{189}$$

We further assume that the concentration is invariant during diffusion, this means that the net amount of impurities that diffused out from the gate polysilicon is much smaller than the initial total amount of impurities in the gate polysilicon, which is always a good approximation.

The diffusion equations in gate oxide and Si substrates are given by

$$\frac{\partial N_I(x,t)}{\partial t} = D_I \frac{\partial^2 N_I(x,t)}{\partial^2 x_I} \tag{190}$$

$$\frac{\partial N_S(x,t)}{\partial t} = D_S \frac{\partial^2 N_S(x,t)}{\partial^2 x_S} \tag{191}$$

where D denotes the diffusion coefficient, and subscripts I and S denote gate oxide and Si substrate, respectively.

The boundary conditions at either interface of the gate oxide are given by

$$m_p N_I(0,t) = N_0 \tag{192}$$

$$m_c N_I(d_I,t) = N_S(0,t) \tag{193}$$

$$D_I \frac{\partial N_I(x_I,t)}{\partial x_I}\bigg|_{x_I=d_I} = D_S \frac{\partial N_S(x_S,t)}{\partial x_S}\bigg|_{x_S=0} \tag{194}$$

$$N_S(\infty,t) = 0 \tag{195}$$

where m_p is the segregation coefficient of B in polysilicon/SiO$_2$ system, and m_c is the segregation coefficient of B in Si/SiO$_2$ system. B segregation coefficients are assumed to be same and is 0.375 [12] in this analysis.

The initial conditions are given by

$$N_I(x,0) = 0 \tag{196}$$

$$N_S(x,0) = 0 \tag{197}$$

Performing the Laplace transformation of Eq. 184 with respect to t, we obtain

$$s\mathscr{L}\{N_I\} = D_I \frac{d^2 \mathscr{L}\{N_I\}}{d^2 x_I} \tag{198}$$

which is solved as

$$\mathscr{L}\{N_I\} = A(s)\exp\left(-\sqrt{\frac{s}{D_I}}x_I\right) + B(s)\exp\left(\sqrt{\frac{s}{D_I}}x_I\right) \tag{199}$$

Performing the Laplace transformation of Eq. 190 with respect to t, we obtain

$$\frac{N_0}{sm_p} = \mathscr{L}\{N_I(0,t)\} \tag{200}$$
$$= A(s) + B(s)$$

Eliminating $B(s)$ from Eq. 199 using Eq. 200, we obtain

$$\mathscr{L}\{N_I\} = \frac{N_0}{sm_p}\exp\left(\sqrt{\frac{s}{D_I}}x_I\right) - A(s)\left[\exp\left(\sqrt{\frac{s}{D_I}}x_I\right) - \exp\left(-\sqrt{\frac{s}{D_I}}x_I\right)\right] \tag{201}$$

Performing the Laplace transformation of Eq. 191 with respect to t, we obtain

$$s\mathscr{L}\{N_S(x,t)\} = D_S\frac{d^2\mathscr{L}\{N_S(x,t)\}}{d^2x_S} \tag{202}$$

which is solved as

$$\mathscr{L}\{N_S\} = C(s)\exp\left(-\sqrt{\frac{s}{D_I}}x_I\right) \tag{203}$$

We neglect the term associated with $\exp\left(\sqrt{s/D_I}\,x_I\right)$ utilizing the boundary condition of Eq. 195.

Performing the Laplace transformation of Eq. 193 with respect to t, and substituting Eqs. 201 and Eq. 203, we obtain

$$\mathscr{L}\{N_s\} = m_c\left\{\frac{N_0}{s}e^F - A(s)\left(e^F - e^{-F}\right)\right\}\exp\left(-\sqrt{\frac{s}{D_s}}x_s\right) \tag{204}$$

where F is defined as

$$F \equiv \sqrt{\frac{s}{D_I}} d_I \tag{205}$$

Performing the Laplace transformation of Eq. 194 with respect to t, and substituting Eqs. 201 and Eq. 204, we obtain

$$A(s) = \frac{\dfrac{(\gamma + m_S) N_0}{s m_p} e^F}{m_c \left(e^F - e^{-F}\right) + \gamma \left(e^F + e^{-F}\right)} \tag{206}$$

where γ is defined as

$$\gamma \equiv \sqrt{\frac{D_I}{D_S}} \tag{207}$$

Substituting Eq. 206 into Eq. 204, we obtain

$$
\begin{aligned}
\mathscr{L}\{N_s\} \\
&= m_c \left\{ \frac{N_0}{s m_p} e^F - \frac{\dfrac{(\gamma + m_c) N_0}{s m_p} e^F}{m_c \left(e^F - e^{-F}\right) + \gamma \left(e^F + e^{-F}\right)} \left(e^F - e^{-F}\right) \right\} \exp\left(-\sqrt{\frac{s}{D_s}} x_s\right) \\
&= \frac{N_0}{s} \frac{m_c}{m_p} e^F \frac{m_c \left(e^F - e^{-F}\right) + \gamma \left(e^F + e^{-F}\right) - (\gamma + m_c)\left(e^F - e^{-F}\right)}{m_c \left(e^F - e^{-F}\right) + \gamma \left(e^F + e^{-F}\right)} \exp\left(-\sqrt{\frac{s}{D_s}} x_s\right) \\
&= \frac{N_0}{s} \frac{m_c}{m_p} \frac{2\gamma}{(m_c + \gamma) e^F - (m_c - \gamma) e^{-F}} \exp\left(-\sqrt{\frac{s}{D_s}} x_s\right) \\
&= \frac{N_0}{s} \frac{2\gamma}{(m_c + \gamma)} \frac{m_c}{m_p} \frac{e^{-F}}{1 - \alpha\, e^{-2F}} \exp\left(-\sqrt{\frac{s}{D_s}} x_s\right) \\
&= \frac{N_0}{s} \frac{2\gamma}{(m_c + \gamma)} \frac{m_c}{m_p} e^{-F} \sum_{k=0}^{\infty} \left(\alpha\, e^{-2F}\right)^k \exp\left(-\sqrt{\frac{s}{D_s}} x_s\right) \\
&= \frac{N_0}{s} \frac{2\gamma}{(m_c + \gamma)} \frac{m_c}{m_p} \sum_{k=0}^{\infty} \alpha^k \exp\left[-(2k+1)\sqrt{\frac{s}{D_I}} d_I - \sqrt{\frac{s}{D_s}} x_s\right]
\end{aligned} \tag{208}
$$

where α is defined as

$$\alpha \equiv \frac{m_c - \gamma}{m_c + \gamma} \tag{209}$$

The inverse Laplace transformation of Eq. 208 gives

$$N_s(x,t) = \frac{2\gamma}{m_c + \gamma} \frac{m_c}{m_p} N_0 \sum_{k=0}^{\infty} \alpha^k erfc \left[-\frac{(2k+1)d_I + \gamma x_s}{2\sqrt{D_I t}} \right] \tag{210}$$

Therefore, the SiO$_2$/Si interface concentration on the Si side is given by

$$N_s(0,t) = \frac{2\gamma}{m_c + \gamma} \frac{m_c}{m_p} N_0 \sum_{k=0}^{\infty} \alpha^k erfc \left[-\frac{(2k+1)d_I}{2\sqrt{D_I t}} \right] \tag{211}$$

If we consider the onset point for gate penetration, $N_s(0,t)$ is much smaller than N_0, and we can approximate that only the first term with $k=0$ dominates the right side of Eq. 211, and reduce it to

$$N_s(0,t) = \frac{2\gamma}{m_c + \gamma} \frac{m_c}{m_p} N_0 erfc \left(-\frac{d_I}{2\sqrt{D_I t}} \right) \tag{212}$$

We can define the penetration as such as N_s becomes a certain critical concentration of N_c, and the corresponding critical time t_2 can be expressed by

$$t_2 = \frac{d_I^2}{4D_I} \frac{1}{\left[erfc^{-1} \left(\frac{m_c + \gamma}{2\gamma} \frac{m_p}{m_c} \frac{N_c}{N_0} \right) \right]^2} \tag{213}$$

We need D_I to evaluate t_2, which are intensively done in [15], and it depends on whether we use B or BF$_2$ for ion implantation species, and are given by

$$D_I(BF_2) = 3.96 \times 10^{-2} \exp \left[-\frac{3.65(eV)}{k_B T} \right] cm^2 \Big/ s \tag{214}$$

$$D_I(B) = 1.83 \times 10^{-2} \exp\left[-\frac{3.82\,(eV)}{k_B T} \right] cm^2\!\!\Big/\!s \qquad (215)$$

Consequently, we obtain the thermal budget for B doped polysilicon gate as

$$t_1 \leq t \leq t_2 \qquad (216)$$

which is shown in Fig. **18**, where $r = 0.9$, and $N_c = {}^{N_0}\!\!\big/\!_{10000}$.

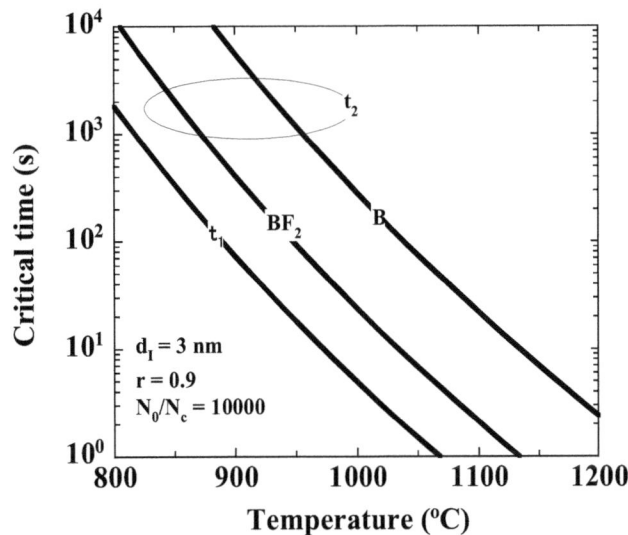

Figure 18: Dependence of critical time on temperature. t_1 correspond to the critical time to make the profile in polysilicon gate flat, and t_2 corresponds to the critical time for gate insulator penetration.

DIFFUSION THROUGH DOUBLE GATE INSULATOR LAYERS

Nitrided-oxide gate was used to suppress B penetration through thin gate insulator layer. Most nitrided oxide contain non uniformly distribute nitrogen with peak concentration at the interface between nitrided oxide and Si substrate. This nitride oxide can be approximately treated as the two layers of SiO_2 and SiON, and the related model was described [15].

The axis for the system is shown in Fig. **19**, where the subscript 1 denotes SiO_2, 2 denotes nitrided-oxide, and 3 denotes Si substrate.

The concentration in gate polysilicon is assumed to be constant and is N_0

The diffusion equations in each layer are given by

$$\frac{\partial N_1(x,t)}{\partial t} = D_1 \frac{\partial^2 N_1(x,t)}{\partial^2 x_1} \tag{217}$$

$$\frac{\partial N_2(x,t)}{\partial t} = D_2 \frac{\partial^2 N_2(x,t)}{\partial^2 x_2} \tag{218}$$

$$\frac{\partial N_3(x,t)}{\partial t} = D_3 \frac{\partial^2 N_3(x,t)}{\partial^2 x_3} \tag{219}$$

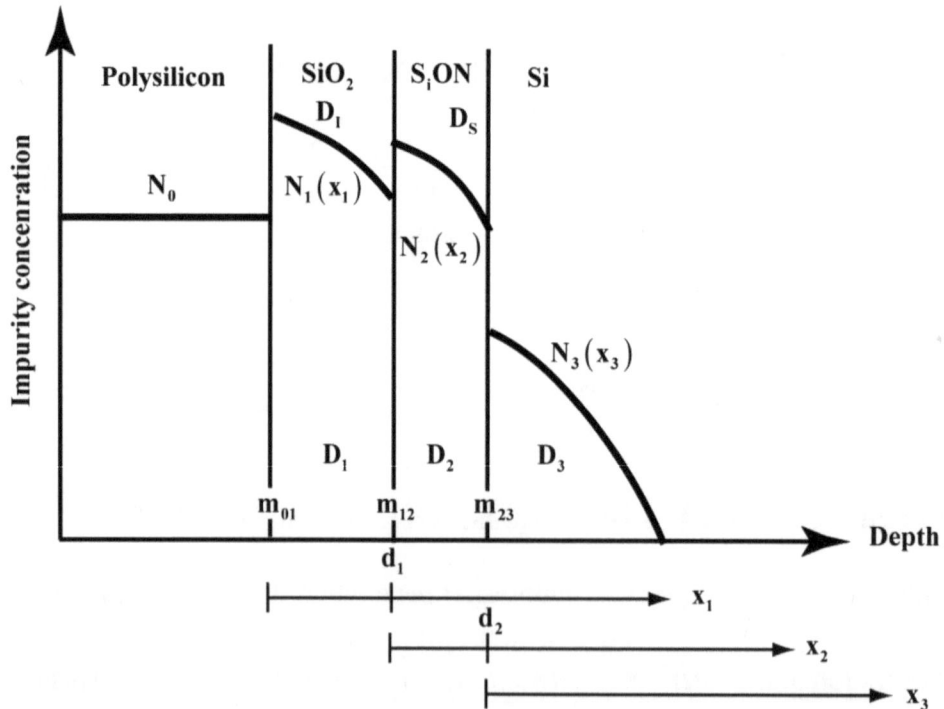

Figure 19: An axis system for impurity diffusion profiles in polysilicon gate/SiO_2/SiON/Si substrate structure. The subscript 1 denote SiO_2 layer, 2 denotes SiON layer, and 3 denotes Si substrate.

We assume that the thermal equilibrium associated the segregation is established. The related boundary conditions associated with concentrations are given by

$$m_{01}N_0 = N_1(0,t) \tag{220}$$

$$m_{12}N_1(d_1,t) = N_2(0,t) \tag{221}$$

$$m_{23}N_2(d_2,t) = N_3(0,t) \tag{222}$$

$$N_3(\infty,t) = 0 \tag{223}$$

where m_{ij} is the segregation coefficient between i and j layers.

Imposing that the flux is continuous, we describe the boundary conditions associated with fluxes as

$$D_1 \left. \frac{\partial N_1(x_1,t)}{\partial x_1} \right|_{x_1=d_1} = D_2 \left. \frac{\partial N_2(x_2,t)}{\partial x_2} \right|_{x_2=0} \tag{224}$$

$$D_2 \left. \frac{\partial N_2(x_2,t)}{\partial x_2} \right|_{x_2=d_2} = D_3 \left. \frac{\partial N_3(x_3,t)}{\partial x_3} \right|_{x_3=0} \tag{225}$$

The initial conditions are

$$N_1(x_1,0) = N_2(x_2,0) = N_3(x_3,0) = 0 \tag{226}$$

Performing the similar Laplace transformation analysis to the case of single gate insulator layer, we obtain

$$\mathscr{L}\{N_1\} = \frac{m_{01}N_0}{s} \exp\left(-\sqrt{\frac{s}{D_1}}x_1\right) - 2A(s)\sinh\left(\sqrt{\frac{s}{D_1}}x_1\right) \tag{227}$$

$$\mathscr{L}\{N_2\} = C(s)\exp\left(-\sqrt{\frac{s}{D_2}}x_2\right) + D(s)\exp\left(-\sqrt{\frac{s}{D_2}}x_2\right) \tag{228}$$

$$\mathscr{L}\{N_3\} = E(s)\exp\left(-\sqrt{\frac{s}{D_3}}x_3\right) \tag{229}$$

where

$$A(s) = \frac{\dfrac{m_{01}N_0}{s}\{1+\eta\alpha\exp(-2\Omega_1)\}}{\{1+\eta\alpha\exp(-2\Omega_2)\}\exp(\Omega_1)-\{1+\alpha\exp(-2\Omega_2)\}\exp(-\Omega_1)} \tag{230}$$

$$C(s) = \frac{m_{12}-\gamma_{12}}{2}\left[\frac{m_{01}N_0}{s}\exp(\Omega_1)-A(s)\left\{\exp(\Omega_1)-\frac{1}{\eta}\exp(-\Omega_1)\right\}\right] \tag{231}$$

$$D(s) = \frac{m_{12}+\gamma_{12}}{2}\left[\frac{m_{01}N_0}{s}\exp(\Omega_1)-A(s)\left\{\exp(\Omega_1)-\frac{1}{\eta}\exp(-\Omega_1)\right\}\right] \tag{232}$$

$$E(s)$$

$$= m_{03}(1-\eta)(1-\alpha)\frac{\dfrac{1}{1+\eta\alpha\exp(-2\Omega_2)}}{1-\dfrac{\eta+\alpha\exp(-2\Omega_2)}{1+\eta\alpha\exp(-2\Omega_2)}\exp(-2\Omega_1)}\frac{N_0}{s}\exp(-\Omega_1-\Omega_2) \tag{233}$$

The variables below are defined as

$$\Omega_1 \equiv \sqrt{\frac{s}{D_1}}d_1 \tag{234}$$

$$\Omega_2 \equiv \sqrt{\frac{s}{D_2}}d_2 \tag{235}$$

$$\gamma_{12} \equiv \sqrt{\frac{D_1}{D_2}} \tag{236}$$

$$\gamma_{23} \equiv \sqrt{\frac{D_2}{D_3}} \tag{237}$$

$$\eta \equiv \frac{m_{12} - \gamma_{12}}{m_{12} + \gamma_{12}} \tag{238}$$

$$\alpha \equiv \frac{m_{23} - \gamma_{23}}{m_{23} + \gamma_{23}} \tag{239}$$

$$m_{03} \equiv m_{01} m_{12} m_{23} \tag{240}$$

Let us consider some special cases.

(a) $D_1 = D_2, m_{12} = 1$

This corresponds to the single gate insulator. $\gamma_{12} = 1$, and hence $\eta = 0$ in this case. $E(s)$ is then reduced to

$$
\begin{aligned}
E(s) &= m_{01} m_{12} m_{23} (1 - \alpha) \frac{N_0}{s} \frac{\exp\left[-(\Omega_1 + \Omega_2)\right]}{1 - \alpha \exp\left[-2(\Omega_1 + \Omega_2)\right]} \\
&= m_{01} m_{12} m_{23} (1 - \alpha) \frac{N_0}{s} \frac{\exp\left[-\sqrt{\dfrac{s}{D_I}}(d_1 + d_2)\right]}{1 - \alpha \exp\left[-2\sqrt{\dfrac{s}{D_I}}(d_1 + d_2)\right]}
\end{aligned}
\tag{241}
$$

We can regard

$$d_I \equiv d_1 + d_2 \tag{242}$$

and

$$m_{01} m_{12} m_{23} = \frac{m_c}{m_p} \tag{243}$$

Therefore, Eq. 241 is reduced to the one for single gate insulator as is expected.

(b) $D_1 \gg D_2$

This is the case when we use oxynitride. We can expect

$$
\eta = \frac{m_{12} - \gamma_{12}}{m_{12} + \gamma_{12}}
$$

$$
\approx \frac{-\gamma_{12}}{\gamma_{12}}
$$

$$
= -1
$$

(244)

We can then simplify Eq. 241 as

$$
E(s) = 2m_{03}(1-\alpha)\frac{1}{1+\exp(-2\Omega_1)}\frac{1}{1-\alpha\exp(-2\Omega_2)}\frac{N_0}{s}\exp(-\Omega_1-\Omega_2)
$$

$$
= 2m_{03}(1-\alpha)\sum_{l=0}^{\infty}\left(-e^{-2\Omega_1}\right)^l \times \sum_{k=0}^{\infty}\left(\alpha e^{-2\Omega_2}\right)^k \frac{N_0}{s}\exp(-\Omega_1-\Omega_2)
$$

(245)

$$
= 2m_{03}(1-\alpha)\sum_{l=0}^{\infty}\sum_{k=0}^{\infty}(-1)^l\alpha^k\frac{N_0}{s}\exp\left[-(2l+1)\Omega_1-(2k+1)\Omega_2\right]
$$

Therefore, we obtain

$$
L\{N_3\}
$$

$$
= 2m_{03}(1-\alpha)\sum_{l=0}^{\infty}\sum_{k=0}^{\infty}(-1)^l\alpha^k\frac{N_0}{s}\exp\left[-(2l+1)\Omega_1-(2k+1)\Omega_2-\sqrt{\frac{s}{D_3}}x_3\right]
$$

(246)

The inverse Laplace transformation of Eq. 246 gives

$$
N_3(x,t)
$$

$$
= 2m_{03}(1-\alpha)N_0\sum_{l=0}^{\infty}\sum_{k=0}^{\infty}(-1)^l\alpha^k erfc\left[\frac{(2l+1)d_1}{2\sqrt{D_1 t}}+\frac{(2k+1)d_2}{2\sqrt{D_2 t}}+\frac{x_3}{2\sqrt{D_3 t}}\right]
$$

(247)

We consider the case where oxyniteride film well blocks the B penetration, and consider the term $k = 0$, and simplify Eq. 247 as

$$
N_3(x,t)
$$

$$
= 2m_{03}(1-\alpha)N_0\sum_{l=0}^{\infty}(-1)^l erfc\left[\frac{(2l+1)d_1}{2\sqrt{D_1 t}}+\frac{d_2}{2\sqrt{D_2 t}}+\frac{x_3}{2\sqrt{D_3 t}}\right]
$$

(248)

We cannot select a certain term associated with l since we cannot assume $(2l+1)d_1 / (2\sqrt{D_1 t})$ is large.

We can assume that on the onset point of B penetration, we can assume that the concentration in SiO_2 is high enough, and hence that we can assume that $d_1 / (2\sqrt{D_1 t})$ is small. We form a pair of l and $l+1$, where l is an even number given by

$$erfc\left(F + \frac{ld_1}{\sqrt{D_1 t}}\right) - erfc\left(F + \frac{(l+1)d_1}{\sqrt{D_1 t}}\right)$$

$$= erfc\left(F + \frac{ld_1}{\sqrt{D_1 t}}\right) - erfc\left(F + \frac{ld_1}{\sqrt{D_1 t}} + \frac{d_1}{\sqrt{D_1 t}}\right)$$

$$\approx erfc\left(F + \frac{ld_1}{\sqrt{D_1 t}}\right) - \left\{erfc\left(F + \frac{ld_1}{\sqrt{D_1 t}}\right) - \frac{2}{\sqrt{\pi}}\exp\left[-\left(F + \frac{ld_1}{\sqrt{D_1 t}}\right)^2\right]\frac{d_1}{\sqrt{D_1 t}}\right\}$$

$$= \frac{2}{\sqrt{\pi}}\exp\left[-\left(F + \frac{ld_1}{\sqrt{D_1 t}}\right)^2\right]\frac{d_1}{\sqrt{D_1 t}}$$

(249)

where

$$F \equiv \frac{d_1}{2\sqrt{D_1 t}} + \frac{d_2}{2\sqrt{D_2 t}} + \frac{x_3}{2\sqrt{D_3 t}}$$

(250)

Therefore, the sum of series associated with l is given by

$$\sum_{l=0,2,4,\cdots}^{\infty} \frac{2}{\sqrt{\pi}}\exp\left[-\left(F + \frac{ld_1}{\sqrt{D_1 t}}\right)^2\right]\frac{d_1}{\sqrt{D_1 t}}$$

$$= \frac{1}{2}\left\{\sum_{p=0,1,2,\cdots}^{\infty} \frac{2}{\sqrt{\pi}}\exp\left[-\left(F + \frac{2pd_1}{\sqrt{D_1 t}}\right)^2\right]\frac{2d_1}{\sqrt{D_1 t}}\right\}$$

(251)

$$= \frac{1}{2}\left\{\sum_{p=0,1,2,\cdots}^{\infty} \frac{2}{\sqrt{\pi}}\exp\left[-\left(F + p\Delta y\right)^2\right]\Delta y\right\}$$

where

$$\Delta y \equiv \frac{2d_1}{\sqrt{D_1 t}} \tag{252}$$

This can be regarded as continuous integration as

$$\frac{1}{2}\left\{ \sum_{p=0,1,2,\cdots}^{\infty} \frac{2}{\sqrt{\pi}} \exp\left[-(F+p\Delta y)^2\right]\Delta y \right\} \approx \frac{1}{2}\int_0^{\infty} \frac{2}{\sqrt{\pi}} \exp\left[-(F+y)^2\right]dy$$

$$= \frac{1}{2}\int_F^{\infty} \frac{2}{\sqrt{\pi}} \exp\left(-z^2\right)dz \tag{253}$$

where

$$z \equiv F + y \tag{254}$$

Therefore, the diffusion profile in Si substrate is given by

$$N_3(x,t) = m_{03}(1-\alpha)N_0 erfc\left(\frac{d_1}{2\sqrt{D_1 t}} + \frac{d_2}{2\sqrt{D_2 t}} + \frac{x_3}{2\sqrt{D_3 t}} \right) \tag{255}$$

We can easily evaluate the effectiveness of the SiON layer as an effective diffusion coefficient D_{eff}

$$\frac{d_1+d_2}{2\sqrt{D_{eff} t}} = \frac{d_1}{2\sqrt{D_1 t}} + \frac{d_2}{2\sqrt{D_2 t}} \tag{256}$$

This leads to

$$D_{eff} = \left(\frac{d_1+d_2}{d_1+\gamma_{12}d_2} \right)^2 D_1 \tag{257}$$

Since $\gamma_{12} \gg 1$, the penetration is significantly suppressed.

REFERENCES

[1]   J. Crank, The Mathematics of Diffusion, Oxford science publication, 1975, New York.

[2]   H. S. Carslaw and J. C. Jaeger, Conduction of Heat in Solid, Oxford science publication, 1975, New York.

[3]   Y. Nakajima, S. Ohkawa, and Y. Fukukawa, "Simplified expression for the distribution of diffused impurity," Japan J. Appl. Phys., vol. 10, pp. 162-163, 1971.

[4]   R. B. Fair and J. C. C. Tsai, "The diffusion of ion-implanted arsenic in silicon," J. Electrochem. Society, vol. 122, pp. 1689-1696, 1975.

[5]   R. B. Fair and J. C. C. Tsai, "Profile parameters of implanted-diffused arsenic layers in silicon," J. Electrochem. Society, vol. 123, pp. 583-586, 1976.

[6]   R. B. Fair, "Boron diffusion in silicon-concentration and orientation dependence, background effects, and profile estimation," J. Electrochem. Society, vol. 122, pp. 800-805, 1975.

[7]   D. Anderson and M. Liasak "Approximate solutions of some nonlinear diffusion equations," Physical Review A, vol. 22, pp. 2761-2768, 1980.

[8]   D. Anderson and K. O. Jeppson, "Nonlinear two-step diffusion in semiconductors," J. Electrochem. Society, vol. 131, pp. 2675-2679, 1984.

[9]   K. Suzuki, "High concentration impurity diffusion model," Solid-State Electronics, vol. 44, pp. 457-463, 2000.

[10]   K, Suzuki, "High concentration diffusion profiles of low energy ion implanted B, As, and BF2 in bulk silicon," Solid-State Electronics, vol. 45, pp. 1747-1751, 2001.

[11]   K. Suzuki and H. Tashiro, "Analytical model for spike annealed diffusion profiles of low-energy and high-dose ion implanted impurities," Rapid Thermal Processing for Future Semiconductor Devices, Elsevier Science, pp. 9-16, 2003.

[12]   ISE TCAD Manuals vol. 2, release 5, Part 7.

[13]   FabMeister IM http://www.mizuho-ir.co.jp/science/ion/index.html

[14]   K. Suzuki, A. Satoh, T. Aoyama, I. Namura, F. Inoue, Y. Kataoka, Y. Tada, and T. Sugii,"Thermal budget for fabricating a dual gate deep submicron CMOS with thin pure gate oxide," Jpn. J. Appl. Phys., vol. 35, pp. 1496-1502, 1996.

[15]   T. Aoyama, S. Ohkubo, H. Tashiro, Y. Tada, K. Suzuki, and K. Horiuchi, "Boron diffusion in nitride-oxide gate dielectrics leading to high suppression of boron penetration in p-MOSFETs," Jpn. J. Appl. Phys., Vol. 37, pp. 1244-1250, 1998.

APPENDIX

A. FOURIER TRANSFORMATION

We utilize Fourier transformation to solve the differential equation.

The basics of Fourier transformation are given by

$$\mathscr{F}\{f(x)\} = \mathscr{F}(\xi) = \frac{1}{\sqrt{2\pi}} \int_{-\infty}^{\infty} f(x) e^{-i\xi x} dx \tag{A-1}$$

$$\mathscr{F}^{-1}\{\mathscr{F}(\xi)\} = \frac{1}{\sqrt{2\pi}} \int_{-\infty}^{\infty} \mathscr{F}(\xi) e^{i\xi x} d\xi \tag{A-2}$$

$$
\begin{aligned}
\mathscr{F}\{f_x\} &= \frac{1}{\sqrt{2\pi}} \int_{-\infty}^{\infty} \frac{df(x)}{dx} e^{-i\xi x} dx \\
&= \frac{1}{\sqrt{2\pi}} \left\{ \left[f(x) e^{-i\xi x} \right]_{-\infty}^{\infty} + i\xi \int_{-\infty}^{\infty} f(x) e^{-i\xi x} dx \right\} \\
&= i\xi \mathscr{F}\{f(x)\}
\end{aligned} \tag{A-3}
$$

$$
\begin{aligned}
\mathscr{F}\{f_{xx}\} &= i\xi \mathscr{F}\{f_x\} \\
&= -\xi^2 \mathscr{F}\{f\}
\end{aligned} \tag{A-4}
$$

The convolution integral is very important and is given by

$$\mathscr{F}\{f*g\} = \mathscr{F}\{f\} \mathscr{F}\{g\} \tag{A-5}$$

where

$$f*g = \frac{1}{\sqrt{2\pi}} \int_{-\infty}^{\infty} f(x-\eta) g(\eta) d\eta \tag{A-6}$$

This is proved as follows:

$$
\begin{aligned}
\mathscr{F}\{f*g\} &= \mathscr{F}\left\{ \frac{1}{\sqrt{2\pi}} \int_{-\infty}^{\infty} f(x-\eta) g(\eta) d\eta \right\} \\
&= \frac{1}{\sqrt{2\pi}} \int_{-\infty}^{\infty} \left[\frac{1}{\sqrt{2\pi}} \int_{-\infty}^{\infty} f(x-\eta) e^{-i\xi x} g(\eta) d\eta \right] dx
\end{aligned} \tag{A-7}
$$

Changing the order of integral, we obtain

$$\mathscr{F}\{f*g\} = \mathscr{F}\left\{\frac{1}{\sqrt{2\pi}}\int_{-\infty}^{\infty}f(x-\eta)g(\eta)d\eta\right\}$$

$$= \frac{1}{\sqrt{2\pi}}\frac{1}{\sqrt{2\pi}}\int_{-\infty}^{\infty}\left[\int_{-\infty}^{\infty}f(x-\eta)e^{-i\xi x}dx\right]g(\eta)d\eta \tag{A-8}$$

Performing the variable transformations of

$$x-\eta = s, \; dx = ds \tag{A-9}$$

we obtain

$$\mathscr{F}\{f*g\} = \frac{1}{\sqrt{2\pi}}\frac{1}{\sqrt{2\pi}}\int_{-\infty}^{\infty}\left[\int_{-\infty}^{\infty}f(s)e^{-i\xi(s+\eta)}ds\right]g(\eta)d\eta$$

$$= \frac{1}{\sqrt{2\pi}}\frac{1}{\sqrt{2\pi}}\int_{-\infty}^{\infty}\left[\int_{-\infty}^{\infty}f(s)e^{-i\xi s}ds\right]g(\eta)e^{-i\xi\eta}d\eta \tag{A-10}$$

$$= \frac{1}{\sqrt{2\pi}}\int_{-\infty}^{\infty}f(s)e^{-i\xi s}ds\int_{-\infty}^{\infty}g(\eta)e^{-i\xi\eta}d\eta$$

$$= \mathscr{F}\{f\}\mathscr{F}\{g\}$$

Table **A-1** shows some Fourier transform pairs.

Table A-1: Table of Fourier tranform pairs

$f(x)$	$F(\xi) = \mathscr{F}\{f(x)\}$
$f(ax) \quad a > 0$	$\dfrac{1}{a} F\left(\dfrac{\xi}{a}\right)$
$f(x-a)$	$e^{-ia\xi} F(\xi)$
$e^{-a^2 x^2}$	$\dfrac{1}{\sqrt{2}a} \exp\left(-\dfrac{\xi^2}{4a^2}\right)$
$e^{-a\|x\|}$	$\sqrt{\dfrac{2}{\pi}} \dfrac{a}{a^2 + \xi^2}$
$xe^{-a\|x\|} \quad a > 0$	$-2\sqrt{\dfrac{2}{\pi}} \dfrac{ia\xi}{\left(a^2 + \xi^2\right)^2}$
$\delta(x-a)$	$e^{-ia\xi} \dfrac{1}{\sqrt{2\pi}}$
$\dfrac{a}{x^2 + a^2}$	$\sqrt{\dfrac{\pi}{2}} e^{-a\|\xi\|}$

B. LAPLACE TRANSFORMATION

We utilize Laplace transformation to solve the differential equation.

The basics of Laplace transformation are given by

$$\mathscr{L}\{f(t)\} = \int_0^\infty f(t)\exp(-st)\,dt \tag{B-1}$$

$$\begin{aligned}
\mathscr{L}\left\{\frac{\partial f(t)}{\partial t}\right\} &= \int_0^\infty \frac{\partial f(t)}{\partial t}\exp(-st)\,dt \\
&= \left[f(t)\exp(-st)\right]_0^\infty + s\int_0^\infty f(t)\exp(-st)\,dt \\
&= s\mathscr{L}\{f(t)\} - f(0)
\end{aligned} \tag{B-2}$$

$$\begin{aligned}
\mathscr{L}\left\{\frac{\partial^2 f(t)}{\partial t^2}\right\} &= \int_0^\infty \frac{\partial^2 f(t)}{\partial t^2}\exp(-st)\,dt \\
&= \left[\frac{\partial f(t)}{\partial t}\exp(-st)\right]_0^\infty + s\int_0^\infty \frac{\partial f(t)}{\partial t}\exp(-st)\,dt \\
&= s\mathscr{L}\left\{\frac{\partial f(t)}{\partial t}\right\} - \left.\frac{\partial f}{\partial t}\right|_{t=0} \\
&= s\left[s\mathscr{L}\{f(t)\} - f(0)\right] - \left.\frac{\partial f}{\partial t}\right|_{t=0} \\
&= s^2\mathscr{L}\{f(t)\} - sf(0) - \left.\frac{\partial f}{\partial t}\right|_{t=0}
\end{aligned} \tag{B-3}$$

The convolution integral is very important and is given by

$$\mathscr{L}\{f*g\} = \mathscr{L}\{f\}\mathscr{L}\{g\} \tag{B-4}$$

whcrc

$$\begin{aligned}
f*g &= \int_0^\infty f(t-\lambda)g(\lambda)\,d\lambda \\
&= \int_0^\infty f(\lambda)g(t-\lambda)\,d\lambda
\end{aligned} \tag{B-5}$$

This is proved as follows.

$$\mathcal{L}\{f * g\} = \mathcal{L}\left\{\int_0^\infty f(t-\lambda)g(\lambda)d\lambda\right\}$$
$$= \int_0^\infty \left[\int_0^\infty f(t-\lambda)g(\lambda)d\lambda\right]e^{-st}dt \tag{B-6}$$

Introducing a step function given by

$$u(t-\lambda) = \begin{cases} 1 & \text{for } \lambda < t \\ 0 & \text{for } \lambda > t \end{cases} \tag{B-7}$$

we obtain

$$f(t-\lambda)g(\lambda)u(t-\lambda) = \begin{cases} f(t-\lambda)g(\lambda) & \text{for } \lambda < t \\ 0 & \text{for } \lambda > t \end{cases} \tag{B-8}$$

Utilizing this function, we change the upper limit of inside integral to infinity, and obtain

$$\mathcal{L}\{f * g\} = \int_0^\infty \left[\int_0^\infty f(t-\lambda)g(\lambda)u(t-\lambda)d\lambda\right]e^{-st}dt$$
$$= \int_0^\infty g(\lambda)\left[\int_0^\infty f(t-\lambda)u(t-\lambda)e^{-st}dt\right]d\lambda \tag{B-9}$$

Let us consider the inside integral. We obtain

$$f(t-\lambda)u(t-\lambda) = \begin{cases} f(t-\lambda) & \text{for } t < \lambda \\ 0 & \text{for } t < \lambda \end{cases} \tag{B-10}$$

Therefore, we can regard the lower limit of integral as λ instead of 0, and obtain

$$\mathcal{L}\{f * g\} = \int_0^\infty g(\lambda)\left[\int_\lambda^\infty f(t-\lambda)e^{-st}dt\right]d\lambda \tag{B-11}$$

Performing a variable transformation of

$$\tau = t - \lambda \tag{B-12}$$

we obtain

$$
\begin{aligned}
\mathcal{L}\{f * g\} &= \int_0^\infty g(\lambda)\left[\int_0^\infty f(\tau)e^{-s(\tau+\lambda)}d\tau\right]d\lambda \\
&= \int_0^\infty g(\lambda)e^{-s\lambda}d\lambda\int_0^\infty f(\tau)e^{-s\tau}d\tau \\
&= \mathcal{L}\{f\}\mathcal{L}\{g\}
\end{aligned}
\tag{B-13}
$$

Table **B-1** shows some Fourier transform pairs.

Table B-1: Table of Laplace tranform pairs

$f(x)$	$F(s) = \mathscr{L}\{f(t)\}$
1	$\dfrac{1}{s}$
e^{at}	$\dfrac{1}{s-a}$ $s > a$
$\sin(at)$	$\dfrac{a}{s^2 + a^2}$ $s > 0$
$\cos(at)$	$\dfrac{s}{s^2 + a^2}$ $s > 0$
$\sinh(at)$	$\dfrac{a}{s^2 - a^2}$ $s > |a|$
$\cosh(at)$	$\dfrac{s}{s^2 - a^2}$ $s > |a|$
$e^{bt}\sinh(at)$	$\dfrac{a}{(s-b)^2 + a^2}$ $s > a$
$e^{bt}\cosh(at)$	$\dfrac{s-b}{(s-b)^2 + a^2}$ $s > a$
t^n	$\dfrac{n!}{s^{n+1}}$ $s > 0$
$erf\left(\dfrac{t}{2a}\right)$	$\dfrac{1}{s}e^{-a\sqrt{s}}$
$e^{at}f(t)$	$F(s-a)$
$f(at)$	$\dfrac{1}{a}F\left(\dfrac{s}{a}\right)$ $a > 0$

INDEX

A

B

C

D

E

F

G

Generation energy 11

H

Henry's law 102

I

Interstitial Si 9
Intrinsic carrier concentration 29
Intrinsic diffusion coefficients 24
Invers Fourier transformation 129
Ionization 38, 41

J

Junction depth 73
Jumping distance 5
Jumping frequency 5

L

Laplace transformation 146, 201-204

M

Massoud model 109
Maximum diffusion concentration 69, 92
Mobility 7
Moments 135

O

Oxidation 101-119

P

Pairing 47
Point defects 9, 35

R

Rapid thermal annealing 57

www.ingramcontent.com/pod-product-compliance
Lightning Source LLC
Chambersburg PA
CBHW050846220326
41598CB00006B/442